Power Up

Power Up

An Engineer's Adventures into Sustainable Energy

YASMIN ALI

hodder
press

First published in Great Britain in 2024 by Hodder Press
An imprint of Hodder & Stoughton Limited
An Hachette UK company

2

Copyright © Yasmin Ali 2024

The right of Yasmin Ali to be identified as the Author of the Work has been asserted by her in accordance with the Copyright, Designs and Patents Act 1988.

All rights reserved. No part of this publication may be reproduced, stored in a retrieval system, or transmitted, in any form or by any means without the prior written permission of the publisher, nor be otherwise circulated in any form of binding or cover other than that in which it is published and without a similar condition being imposed on the subsequent purchaser.

A CIP catalogue record for this title is available from the British Library

Hardback ISBN 9781529382976
Trade Paperback ISBN 9781529382983

Typeset in Celeste by Hewer Text UK Ltd, Edinburgh
Printed and bound in Great Britain by Clays Ltd, Elcograf S.p.A.

Hodder & Stoughton policy is to use papers that are natural, renewable and recyclable products and made from wood grown in sustainable forests. The logging and manufacturing processes are expected to conform to the environmental regulations of the country of origin.

Hodder & Stoughton Limited
Carmelite House
50 Victoria Embankment
London EC4Y 0DZ

www.hodderpress.co.uk

For Ban and Mohammed-Salih

إهداء إلى بان و محمد صالح

Contents

Introduction · · · 1

PART 1: GET IT

Chapter 1: Fossil · · · 17
Chapter 2: Nuclear · · · 43
Chapter 3: Sun · · · 69
Chapter 4: Wind and Water · · · 91

PART 2: MOVE IT

Chapter 5: Sea and Land · · · 117
Chapter 6: Pipelines · · · 137
Chapter 7: Cables · · · 163
Chapter 8: Energy Storage · · · 189

PART 3: USE IT

Chapter 9: Heat and Cool · · · 213
Chapter 10: Industry · · · 237

Chapter 11: Transport	261
Epilogue	285
Further Reading	295
Acknowledgements	297
Endnotes	301
Index	323

Introduction

MY FAMILY'S CAR in Baghdad, Iraq, like many others in the city, was a petrol-powered Volkswagen Passat. It was nationally nicknamed the 'Barazilli', and its prime feature was that it constantly broke down. It felt as if we spent more time pushing it to get it going again than actually being inside it. The rumour was that a bad batch of Volkswagen Passats manufactured in Brazil (hence the nickname) had made their way to Iraq. For a country with abundant oil and gas resources, the failure of our petrol-powered car, alongside the constant power cuts we experienced, was ironic, to say the least.

Water was also one of many commodities that was not as available as it should have been in the early 1990s in Iraq. A few times a week our house phone would ring. It was a grey, old-style phone with a rotary dial and a coiled cable, and I remember picking it up to hear my aunt's voice on the other end, telling me, 'The water is coming on.' Even at five years old, I knew what to do. I would start filling up jugs of water, before telling my parents that the utility had turned on the water supply. The water supply only came on sporadically, so this was a vital opportunity to get laundry done and fill up the water tanks for drinking, cooking and other daily tasks.

The patchy supply of energy and water makes it sound as if we were living in another era, and certainly not in a country rich in fossil fuels. I blame Iraq's abundant oil resources, and how they were handled, for my parents' decision to eventually leave the country. By the early nineties, having lived through multiple wars, they decided that staying in Iraq was not a viable option if they were to raise a family and educate their children for a successful future. They packed up the essentials, got rid of the Barazilli, bundled me and my siblings into another petrol-powered car and drove to neighbouring Jordan. The twelve-hour, 900-kilometre journey was a very long ride into an unknown future, but with air sanctions in place, it was the only option available.

I was seven years old at the time. This was 1994, and it marked the beginning of an unintended four-year Middle East family tour. This included some time in Benghazi in Libya, a short stint in Tunisia and a few years in Amman, Jordan. After attending six different primary schools, I arrived in the UK with my family. All that moving around at a young age made me comfortable with a nomadic lifestyle, and accustomed to change and uncertainty. It also fired up my sense of curiosity about the world around me. As a result of the impact of fossil fuels on my life, it was impossible not to be fascinated by how these resources that come out of the ground impact individual lives for better and for worse.

After several years of living in the UK, I had forgotten about the struggles of the Barazilli, the patchy water supply and the power cuts – like most people, I took the UK's incredibly reliable energy supply for granted. But the energy industry continued to stitch itself into the fabric of my life – my

INTRODUCTION

early experiences must have had an influence, because I chose to study chemical engineering at university.

It was during a summer job at a gas-fired power station that I really woke up to the enormity of the energy sector and decided I wanted to be part of it. After finishing my chemical engineering degree, I started my career by working in fossil fuels – coal- and gas-fired power stations, followed by oil and gas exploration and production. But a nagging feeling and growing realisation that the future of energy lay elsewhere made me take a leap into the energy transition world. After four years of working in energy innovation for the UK government, funding the development of solutions to help the world reduce greenhouse gases and their impact on the planet, I now work in low-carbon hydrogen project development.

As the years of my career rolled by, more evidence of the necessity to harness, move and use energy unfolded before my eyes. No matter where we live on the planet, energy underpins life as we know it. From the moment we are born, we are exposed to artificial lighting powered by electricity and medical equipment manufactured from materials mined using an energy source. As we grow into toddlers, we play with plastic toys made from oil-derived chemicals, and fall asleep to calming nightlights that project tiny stars onto the ceiling, powered by electricity. As adults, we rely on cars, buses or trains to get us to work, and maintain contact with friends and relatives through telecommunications systems that only work because of reliable energy supplies. With these amazing energy sources we can power equipment to purify water and pump it through a maze of pipework, to keep us hydrated and alive. Soaps, toothpastes and bleaches keep us, our teeth and our homes clean, and again are manufactured only because there is an energy supply to enable it. Nitrogen-based fertilisers, made from energy-intensive ammonia, are spread across fields to provide plants with the nutrients they need to grow quickly and feed us.

We rarely think about the system that props up our existence. But there is a vast global network of power stations, gas pipelines, transmission cables and engineers like me quietly working behind the scenes. They make sure there is enough electricity for the lights to turn on instantly when we flick a switch, and a reliable flow of hot water available for that relaxing bath. We grumble if the heating or air conditioning doesn't work for an hour, underestimating the vast journey society has undertaken to make these services available.

As well as powering our day-to-day lives, the milestones of progress in human civilisation have been marked by our ability to harness and use energy sources. From the very beginning of the human story, we have made use of the sun, which provides the energy needed for plants to grow. We burned wood and organic matter to generate heat for warmth and cooking, as well as manipulating metals into useful tools, propelling society forward, and eventually leading to the Industrial Revolution. Since then, we have become masters at transferring energy from one form to another. Feed in coal, gas or wind at one end and you get electricity at the other, ready to power phones, washing machines, cars and myriad appliances.

It is an impressive journey, but not without mistakes. The way we have used and abused fossil fuels – our society's wastefulness of resources and our desire to own the latest

INTRODUCTION

clothes, cars and gadgets – is releasing greenhouse gases into the atmosphere, causing damage to the environment. If we truly want to understand the challenges we face with climate change, we first need to understand the energy system from the bottom up.

In this book, I hope to show you a complete picture, from the intricate detail of pipelines being welded together to the colossal ships moving vast quantities of liquefied natural gas around the oceans, and from the large-scale energy consumption of our manufacturing industries to the small but powerful actions individuals and communities are taking. We will move through the energy system from start to finish, beginning with the raw resources we can harness for energy (in other words, how we *get* energy), to how we *move* energy around the globe from original source to end user, and finally how we *use* that energy, from transport to food production. However, this book can't cover every single corner of this vast global system, and so the journey we go on will be told through stories of my own adventures, as well as those of others from the past and the present, to give an overall picture of how these feats of science and engineering interact with real lives, including my own.

Getting hold of energy can be done in many different ways. Methods have shifted over the years, and right now we are transitioning from fossil fuels to renewables. Extracting the fossil fuels on which we so heavily rely – coal, oil and gas – has evolved into using drilling rigs and constructing platforms in the middle of the sea, requiring the use of helicopters and boats; an industry with a rich history. Another historically significant energy resource is nuclear power, which started out as a weapon but morphed into a low-emission energy source. Luckily for humans, there is plenty of energy hitting the Earth from the sun; not only is this responsible for the coal, oil and gas reserves we rely on, but also some of our renewable energy sources. Over time, we have learned to harness these resources by converting them into heat, light, movement and electricity.

Once we get these resources and convert them into a useful form, they need to be moved from the point where they are converted to the point where they get used. Oil can be extracted from a well in the Saudi Arabian desert, but it then has to travel through a pipeline to a refinery to be processed, before traversing the ocean in huge oil ships, then being poured into the cylindrical drum of a truck in order to reach a remote petrol station and fill up the pumps. Only then can you fill up a car with petrol. Likewise, natural gas flows through pipelines snaking along the seabed, then criss-crosses its way through overground and underground pipes to get to a gas hob before you can cook a meal.

The electricity that powers our lights and laptops, however, is tricky. Unlike oil and gas, it's not easy to store. This means electricity demand and supply has to be carefully matched, second by second. Storing electricity to use later, by taking advantage of chemical reactions in batteries or in pumped hydropower stations, means more electricity from renewables like wind and solar can be used.

Once the energy is in the form we want it, and we've moved it to where it needs to be, there are many different types of end users. The largest consumers of energy are the heating and cooling of homes, the manufacturing of products and the transporting of people and goods. Huge amounts of energy are used to heat and cool our environments, and as summers get hotter, air conditioning will become more essential – consuming more electricity and potentially exacerbating greenhouse gas emissions. This highlights the urgent need for renewable electricity, and alternative home designs and technologies for cooling. Industry makes the stuff we use, directly and indirectly. For example, fertiliser is needed to grow food – and fertiliser relies on the manufacture of ammonia, which consumes large amounts of energy. Steel, cement and glass are the main ingredients of our built environment, and producing these materials uses up (you guessed it) lots of energy.

Cars, buses, trains, planes and ships guzzle fuel to get us around and to transport the things we want and need from one place to another. The birth of the internal combustion engine is very closely linked to the enormous rise in oil consumption that the globe has seen over the past hundred years or so; it will take a huge cultural shift to change this. Likewise, international travel is becoming a staple of many lives, bringing different people and cultures together, but planes release greenhouse gases, damaging the planet – another area where low-carbon alternatives are urgently needed. We see a similar story in shipping, with fragile island nations like the Marshall Islands being most at risk from climate change yet still reliant upon fossil fuel-powered boats to travel between the islands.

The world is a completely different place today compared to the days of the Industrial Revolution. Nowadays, energy also enables the digital revolution, a major turning point for our society. The internet, vital for everyday activities, relies on an enormous energy supply, as does our use of smartphones and laptops. And in the future, as our lives become increasingly dependent on and intertwined with digital technologies like artificial intelligence, reliable energy supplies will become ever more crucial for everything from medical procedures to security systems.

But beyond these big systemic building blocks and the technical engineering involved, the energy industry is underpinned by people – none of this would be possible without them, and they will be the ones to implement any changes. It is not a faceless monolith, even though that can be how it comes across at times. I am one of those people making this global network a well-oiled machine, allowing you to effortlessly turn on lights, cook your food, travel across the globe, stay warm when it's snowing and cool when it's baking hot, use a computer, and all the countless other conveniences that these days we just couldn't live without.

*

Climate change is primarily an energy problem. It has come about because of the vast quantities of greenhouse gases entering the atmosphere – as a global population, we emit somewhere in the region of 50 billion tonnes of these gases each year, up from less than 10 billion tonnes per year before 1900.[1] Aside from about a fifth of this number, which comes from agriculture, the rest is tied to how we consume energy, so there is a strong link between energy use and greenhouse gas emissions.

INTRODUCTION

Just as my own career has transitioned from fossil fuels to more sustainable sources, as a society we are experiencing a moment of transition for the energy sector as a whole – turning away from fossil fuels and asking what is next. How can we better support our modern way of life without further damaging the planet? How can we move energy from where it is produced to where it is used with a lower carbon footprint? And how is it best to use it at the other end? Complicating matters somewhat, this vital transition is taking place against the backdrop of an energy crisis caused by a complex mix of factors, affecting people across the world. There are deep economic, social and political issues behind this, and I am not aiming to analyse the causes of the recent energy crisis, but I think it is important to understand the energy system itself to help people start to unpick some of these issues. Where does our energy come from? How is it moved around the globe? Who uses it and how? And what are the alternatives for the future?

One of the challenges for the energy transition is the legacy of the existing infrastructure. We are not starting with a blank page and designing a perfect system. We can't knock it down and start from scratch. It is like moving into a new home and making it your own – you can make some changes but you will ultimately be limited by the existing foundations and structures. Because of the existing system, switching is not so simple. It is not a case of turning off all fossil fuels one day and turning on renewables the next. The interconnected nature of the energy system – how an issue on one side of the world can ripple across the globe – adds yet more complexity to this task. But it is possible to work with the existing system, taking advantage of the features, rather than against it.

You might be wondering how we got to this point in the first place, where our lives rely so heavily on such polluting energy sources. Humans knew about fossil fuels as long ago as Ancient Mesopotamia (an area that now includes modern-day Iraq) 4,000 years ago, but it is only since the Industrial Revolution in the 1700s and 1800s that the intensive use of fossil fuels has led us down a path of dependency. Fossil fuels, mainly coal, powered the Industrial Revolution, but were there other options? Events can unfold in many different ways; there were many different routes we could have taken during the twists and turns of the development of the energy system. Perhaps, in a parallel universe, the Industrial Revolution could have been powered by the sun, or even nuclear fusion energy, if certain scientific breakthroughs had happened much earlier. If that had been the case, maybe today we would be dealing with a raft of different challenges other than climate change.

As the energy system grew, so did the fundamentals needed to understand it. When I talk about energy, I mean it in an all-encompassing way. I am thinking about the energy we use to go about our daily lives. This includes electricity, heating, transport and industrial energy uses. In my experience, conversations about energy are often reduced to electricity supplies, but electricity only represents about one-fifth of the energy used globally.

The world of energy is filled with an expansive list of confusing units – kilowatt hours, therms, British Thermal Units, barrels of oil equivalent, and so on. Going back to the most basic definition of energy is a good place to start to untangle these terms. In physics, energy is defined via 'work'

INTRODUCTION

– this is the energy that is transferred when a force acts on an object, moving or displacing it. For example, I pick up an apple, transferring some of the chemical energy I have stored in my body from the food I eat into kinetic energy to take the apple from table height up to my face. The unit for measuring work is the joule, named after the English physicist James Prescott Joule. It is a way of expressing the amount of energy needed to perform a task – if the apple weighs 100 grams, and I want to lift it up by one metre, I will need approximately one joule of energy. If you're making a cup of tea, you need to put in some energy to raise the temperature of the water – about one joule to raise each gram of water by 0.24 degrees Celsius. So, the joule measures the amount of energy.

While measuring energy is useful, it is also important to have a unit to measure power – the rate at which the work is performed. The most fundamental units for this are joules per second, or watts, named after the Scottish inventor James Watt. Think of a tap filling a bath. The rate at which the water is flowing out of the tap into the bath is comparable to the watts, and the amount of water in the bath is comparable to the joules. A watt is a relatively small amount of energy. To put it into perspective, a typical kettle will consume around 3,000 watts, or 3,000 joules every second. There are millions of kettles and other appliances, so the numbers get huge very quickly. A large coal-fired power station will have a capacity of 5,000,000,000 watts. This number is difficult to work with, which is why we divide it by a thousand to get to 5,000,000 kilowatts (kW), and then divide again by a thousand to get to 5,000 megawatts (MW), and by another thousand to get five gigawatts (GW). To put it another way, this power station can generate five gigajoules of energy every second. So how much does it generate in one hour?

This is where the kilowatt hour, the unit commonly found on many electricity and gas bills, comes in. It is made up of the power in kilowatts multiplied by the amount of time in hours. So, if you leave the 3,000-watt (or three-kilowatt) kettle on for an hour, it will consume three kilowatt hours. If the kettle is left running for five hours, it will consume fifteen kilowatt hours, and so on. Returning to the running bath idea, a kilowatt hour is comparable to the volume of water in the bath after running the tap for a defined amount of time. It is useful for engineers to flip back and forth between these units, depending on the tasks at hand and scale needed.

Armed with an understanding of the units, we can now follow the journey of energy, starting with the resources. The most important rule to remember is that energy cannot be created or destroyed, it can only be converted from one form to another. This is the First Law of Thermodynamics, and it comes into play when thinking about the 'primary' energy supply versus energy consumption. The energy contained in a kilogram of coal is the primary energy, but by the time that coal is converted into something useful like electricity, more than half of its energy is wasted, so the final consumption number is lower.

In 2022, the world's primary energy – that is, the energy resources we started with before converting them to other forms – was just over 168,000,000 gigawatt hours.[2] For context, the largest solar farms in the world today, like India's Bhadla Solar Park, can generate somewhere in the

INTRODUCTION

region of 4,000 gigawatt hours in one year. Between primary energy and final consumption by end users, about 30% of that massive number is lost – this is because of inefficiencies in converting one form of energy into another, or inefficiencies in petrol engines in cars, or losses of electricity in the form of heat as it travels along cables. There are many examples of this leaky energy system, but it is well known and scientists and engineers account for it and try to reduce losses as far as possible.

Early on in my career, I was not able to join the dots of the vast global energy systems, which form an organism that crosses borders and connects cultures. While my wider experience in the energy sector means I feel more equipped to do this nowadays, there is still a lot for me to uncover, and much that I discovered when writing this book. I have done my best to include a diverse range of people, places and technologies from across the globe – from the liquified natural gas ship captain in the South China Sea, to Australian citizens embracing rooftop solar, and the battery system powering a home in the wilds of Canada. But these stories are by no means exhaustive – our global energy system is far too vast for me to cover in its entirety. This is a snippet in time and a view backstage of the energy sector at a key moment of change.

My experiences over the past years, the places I have visited, and, most importantly, the people I have met, have made me realise that energy is the basis of our society. We all rely on it every moment of every day. As you read on, I hope you can look over my shoulder as I share these experiences, and see the bigger picture of this hidden world: society's quest for energy resources, and how it has driven our development on this planet. Wherever we are in the world, we all live under one sun, with the same limited resources at our disposal on a global level. How we choose to use these resources will determine the path of our planet and its people into the future.

PART 1

GET IT

Chapter 1

Fossil

I QUICKLY RIPPED off the Velcro. The mouthpiece for the breathing apparatus attached to my flight suit dropped into my hand. I took one last deep breath, and bit down on the mouthpiece. One hand was on my seat belt, and the other on the release mechanism of the helicopter window. I closed my eyes. The helicopter became submerged in water, and rolled, flipping me upside down. Trying my best to keep calm and ignore the disorientation of being upside down and underwater, I unbuckled my seat belt and unlatched the window. I fought against the weight of the water and pushed the window away. Grabbing onto any edges I could find, I felt my way out and ascended to the surface. I had just successfully completed the final and most important task of offshore survival training.

I attended this offshore survival-training course about two years into my career in energy, after taking a job as an operations engineer for an oil and gas exploration and production company. I had worked in coal- and gas-fired power stations, but I was nowhere near understanding the whole picture of global energy. I knew there was more to it than I had seen, so wanted to take a step back from electricity generation to learn how fossil fuels are located, extracted and transported. I was at the start of my energy journey, yet to comprehend the full magnitude of the impact of fossil fuels on the planet, and yet to make the jump from fossil fuels to renewable energy. Despite being on the wrong side of the energy transition during those years, the time I spent in this area did add to my dashboard view of the energy system.

In many parts of the world, survival training is mandatory for anybody visiting an offshore oil and gas platform or rig. Getting offshore usually involves a helicopter ride, so learning how to respond in the worst-case scenario – a crash – is essential. As the heavy machinery of a helicopter sits on top of the vehicle itself, it tends to roll over, lurching its passengers around with it. The best way to learn how to survive one of these incidents is by practising, which is why there are special training centres with fake helicopters attached to cranes suspended above swimming pools.

There's a lot to remember, like how to put on and properly zip up the flight survival suit donned before boarding any helicopter. It's awkward to get into, similar to putting on a dry suit for diving or snorkelling in cold water. Tight, thick elastic bands slip over your feet, wrists and neck, and form a watertight seal. A variety of contraptions are attached to the suit. The breathing apparatus, a small air pouch, gives you a few extra breaths. There's a light, a whistle, and a radio beacon that transmits signals to allow emergency services to locate you. It's uncomfortable and heavy, but it could save your life.

I found the whole experience fine, even easy. But we were doing it in a swimming pool at a pleasant bath-water temperature. I distinctly recall the trainer showing us a video of an

Olympic swimmer swimming in water at around four degrees Celsius. It only took a minute or two for her movement to be visibly impacted by the cold. Her limbs slowed down almost to a halt. I can swim, but I don't expect to compete in the Olympics anytime soon, so the prospect of a helicopter crash into the icy waters of the North Sea terrified me. But to understand fossil fuels, how we extract them and why we rely on them so heavily, I had to complete the course and get myself to an offshore rig to see it all with my own eyes.

*

Most of us are pretty familiar with the origin story of fossil fuels: millions of years ago, some prehistoric trees, plants and marine organisms met their fate. Having lived a full life, soaking up the sun's energy, either directly through photosynthesis or indirectly by eating other plants and creatures, they reached the end of the road. The dead land-based plants ended up at the bottom of bogs and coastal swamps, while the tiny sea creatures – bacteria, plankton and algae – got covered in layers of sand, silt and mud. As millennia went by, more layers of sediment were deposited on top, and turned into rock, permanently trapping the corpses. The rocks weighed down heavily, increasing the pressure and temperature. This cooked the plants and trees into the black rock we know as coal, and the sea creatures into petroleum, the Latin word for 'rock oil'.

Burning fossil fuels, and the associated pollution of our planet, may appear to be a modern phenomenon, coming hand in hand with the Industrial Revolution. But in fact, the use of coal has a long human history; it may have been burned to cremate the dead in Wales during the Bronze Age, the Aztecs in the Americas might have used it for fuel and to make ornaments, and archaeological evidence suggests coal was used in a consistent way as an energy source in China and Inner Mongolia in around 1900 to 2200 BCE (around 4,000 years ago).

Coal has also long been the key to our ability to manipulate metals, which allows us to make useful tools, and build ships and structures. Blacksmiths needed a fuel to generate extremely high temperatures, and coal played a part in the production of iron from iron ore. In Europe, coal was used by soap makers, and it was needed to make plaster from limestone for building. It was also used by salt boilers, who burned it to boil seawater to make salt, which was used to preserve food.

In the 1500s and 1600s in London, the city I now call home, coal was rightly feared to be toxic to health, so people were reluctant to use it for heating and cooking. But the alternative fuel, wood, was in short supply, as Londoners consumed trees faster than they could be replaced. They had no choice but to burn coal to stay warm and cook.[1] Unsurprisingly, this wrapped London in a thick blanket of smoke. John Evelyn, an English writer and diarist, visited London in 1659 and described it as ugly, covered in a cloud of coal smoke, and 'hell upon Earth'. He also highlighted the breathing difficulties and constant coughing of the city's inhabitants: 'I have been in a spacious church where I could not discern the minister for the smoke, nor hear him for the people's barking.'[2] This problem got worse over a long period of time before it got better, when the 'Great Smog of London' in 1952 finally triggered the Clean Air Act. The Act brought in new measures to tackle air pollution, including a shift to smokeless fuels in densely populated areas.[3]

Coal was already an essential fuel for civilisations all over the world long before the invention of the steam engine and the rise of the Industrial Revolution, when people moved from looking after farm animals, growing crops and making things by hand to inventing new machines for manufacturing and working in factories. By the 1900s, huge amounts of coal were being extracted all around the globe in order to meet the vast majority of the world's energy needs.[4] On top of being used to power iron and steel manufacturing, the textiles industry, railways and heating and cooking systems in the home, coal was even used for electricity production. The world's first coal-fired public power station, the Electric Edison Light Station, was built in London in 1882 by Edward Johnson, Thomas Edison's business associate. I have walked up and down the street it was on, but no visible signs of it remain today, as many of the buildings in the area were destroyed during the Second World War. It generated 93 kilowatts of electricity, enough to power about thirty of today's household electric kettles. At the time, it lit street lamps and provided electricity for wealthy private residences.

My first job as an engineering graduate was at a coal-fired power station called Ratcliffe-on-Soar, near the home of Robin Hood in Nottinghamshire. This power station could generate 2,000 megawatts – or, for the sake comparison with the Electric Edison Light Station, it produced 2,000,000 kilowatts, enough electricity for around two million modern homes. On my first day, as I drove past green fields set against a backdrop of blue skies and fluffy white clouds, small grey structures in the distance grew bigger as I approached, revealing the detail of the eight curvaceous concrete cooling towers.

Nothing could have prepared me for the sense of awe I felt when I finally parked my tiny Nissan Micra, dwarfed by these massive structures. I was pleasantly surprised by the sound, too: gushing water. I felt as if I was standing next to a giant waterfall. The name 'cooling tower' describes the function – removing waste heat from the process by passing the steam up the hollow centre of the tower, cooling and condensing it back into liquid water, which flows downwards into the ponds below the towers.

The huge cooling towers and large amounts of heat being removed from the process are an indication of the efficiency of a coal-fired power station, at roughly 30%. This describes how much of the energy contained in the coal is turned into electricity; the remaining 70% of the energy is wasted as heat to the atmosphere. Seventy per cent sounds like a lot of waste, but this is partly dictated by the laws of the universe: thermodynamic laws mean that it is impossible to convert 100% of the energy contained in a fuel to useful energy; some will always be lost to the environment.

A coal-fired power station like Ratcliffe-on-Soar is known as a thermal power station. These crop up very frequently in the world of energy, coming in different shapes and sizes, and operating slightly differently, but all use the same basic principles. Heat energy, in this case from burning coal in a furnace to release the chemical energy it holds, heats water flowing through the pipes of the power station's boiler, turning it to high-pressure steam. The steam squeezes its way over to the steam turbine – imagine a long metal tube with multiple rows of blades fanning out of it. As the steam hits the blades, it forces them to move, rotating the metal tube with them. The turbine is completely enclosed, so the steam has only one pathway in and out, past the blades. The metal tube is physically linked to a generator, so that spins too, and converts the kinetic energy to electrical energy.

While at Ratcliffe-on-Soar, I managed to get a peek into one of the gigantic boilers during an outage – the time of year when certain parts of the power station are turned off for maintenance. The sight above me was not what I was expecting: several people on ropes, abseiling down one of the walls of a boiler several storeys high. Inside the boiler, endless closely packed tubes stretched as far as I could see.

'What are they doing?' I asked my colleague, unable to make any sense of it.

'They're knocking the slag off the walls,' I recall him telling me, making it seem like that was a completely normal thing to be doing.

Over time, burning coal creates molten ash, which builds up on the boiler walls and tubes. This build-up, called 'slag', looks like a very extreme version of the scaling you may get in a kettle if you live in an area with hard water. As is the case for a kettle, it makes the process of transferring the heat from where it is generated to where it is needed more difficult, reducing the efficiency of the power station and wasting even more of the precious energy. The people on ropes were physically removing this stuff, with what I can only imagine was some sort of ice pick designed for the job.

Spending time at this power station and getting into its bowels gave me an appreciation of how far we have come since the Industrial Revolution, and I am grateful for the engineers and technicians who run these places 24/7 to make sure we have an uninterrupted supply of electricity. Coal is an extremely useful energy store, and has served us well, but at a cost. As a fossil fuel, when burned it releases carbon dioxide, nitrous oxides, and sulphur oxides, which contribute to climate change as well as air pollution.

Instead of closing down, however, some coal-fired power stations are adapting. One bright spring day, Fiona Macleod, at the time the managing director of Lynemouth power station on the north-east coast of England, showed me around the site.[5] Fiona is a chemical engineer like me, who has spent over forty years working in the chemicals and energy sectors. She is endlessly knowledgeable, engaging and fun to be around (and she has a side-career as an author, writing chemical engineering-based crime fiction under the pen name Fiona Erskine!). The power station she managed at Lynemouth started life burning coal to generate electricity for a nearby smelter, which extracted aluminium from alumina. Instead of shutting down to meet the UK government's plan to close all coal-fired power stations by 2024, it was converted to burn biomass – fuel that comes from recently living but now dead plants, as opposed to fossil fuels, which come from creatures that died a very long time ago.

Burning biomass to generate electricity is a divisive subject. At the moment, it is generally considered carbon neutral because an equivalent amount of the carbon that is released in energy production is absorbed by the newly planted biomass feedstocks through photosynthesis.[6] But it's not quite that simple; the overall net emissions from using biomass – that is, the amount of carbon dioxide absorbed by the plant minus the amount emitted from power generation – depends on many factors. Where did the biomass come from? Will it be replanted, and if so, with what type of plant, and how will the land be managed? How far has it travelled, and how much carbon dioxide was emitted by the ships and trucks that moved it?[7] Added to that, the amount of land we have here on Earth is finite, and needed for other important things like growing food, so the quantity of biomass available for energy generation will always be limited.[8]

Given all of this, when I first heard of power stations burning biomass to generate electricity, I was sceptical. I immediately began to imagine huge forests being cut down and trees shredded up. But Fiona challenged my understanding, highlighting the strict biomass sustainability criteria that are in place, which mean that Lynemouth burns pelletised forestry and sawmill residues like sawdust, bark and tree thinnings – anything that cannot be turned into useful items like furniture. Key to biomass sustainability are the regulations and complex carbon accounting rules, which are already in place in many areas of the world. Going forward, more decarbonisation initiatives (as countries race to reduce their emissions) will likely drive up demand for biomass, making sustainability criteria even more vital. If these rules are followed and biomass is truly sourced in a responsible and sustainable way, with all of these issues dealt with, it could be a useful part of the future energy mix, alongside other low-carbon sources of energy, and an option for some ageing coal-fired power stations.

Unlike other power stations I have spent time at, Lynemouth radiates an inviting wooden sauna smell. But, very much like the coal-fired power station I worked at as a graduate, the boiler has a set of water-filled tubes, which heat up and turn the water into steam to spin a turbine and generator, generating electricity. Luckily for me, one of the station's three boilers was turned off for maintenance, so I got to pop my head inside and marvel at the tubes. The station's other two boilers were running, but I could peek in through a small port in the boiler wall. I didn't get too close, as I could feel the intense heat and see the metal boiler tubes glowing red, with flames roaring, but I spotted a clear build-up of slag on the tubes – again, very much like a coal-fired boiler.

While some parts of the site had to be changed for the coal-to-biomass conversion, the way the process works remained largely the same. Since the equipment and infrastructure already exist, it may make sense to repurpose it into something more sustainable rather than shutting down and demolishing it (a huge use of energy in its own right) – that is, as long as we can be sure the biomass fuel is sustainably sourced and will not cause more harm than good. The efficiency of the biomass power generation process is similar to that of coal, at around 30%. A small amount of the waste heat is used to warm up a nearby pond used to grow worms for fishing – I could see this pond when we went up on the roof. Beyond it, there were some wind turbines, and a beautiful beach.

Though many coal-fired power stations in the UK and Europe are getting phased out of service or adapting, as of 2022 there were still around 7,000 stations up and running around the world.[9] The use of coal is declining in Europe and the USA, but it is still in high demand in places like India, China and South East Asia, where it is accessible, convenient and cheap, but comes with an environmental cost.[10] Given that Europe and the USA enjoyed many years of using coal and the ensuing economic growth, it seems hypocritical to now preach ending its use to others – we are in the depths of ethically murky waters.

However, given the increasing concern about climate change and the environment in all corners of the world, and the success of renewable energy technologies, perhaps those stepping up their coal use today will start to dial it back. Those who have polluted in the past, instead of judgement, can offer support to leapfrog coal straight to renewable energy. That is certainly my hope. But when it comes to burning fossil fuels, coal is just one choice on the menu.

*

Thousands of years ago, the people of ancient Mesopotamia used fossil fuels, namely oil, but in a very different way to how it is used today, and without causing much damage to the environment. This land, bordered by the Tigris and Euphrates rivers, includes modern-day Iraq, where oil still seeps out of the ground. In ancient times, people used the thick, sticky oil to waterproof boats, as mortar for buildings, and to set jewels and mosaics. Some people even point to the possible use of oil referenced in the Book of Genesis in the Bible, where it is written that 'pitch', an oil product, was used as mortar in the Tower of Babel, and that Noah waterproofed his boat with it. The town of Hit, on the banks of the Euphrates, had seemingly endless supplies of the fossil fuel spewing from the ground.[11]

Fossil fuels are not inherently bad; it all depends on how they are used and the quantities involved. These early uses of oil were in tiny quantities by today's standards, and as a glue it had little impact on the environment, as it did not release carbon dioxide into the atmosphere. The big game-changer in fossil fuel use was the invention and improvement of the internal combustion engine in the late 1800s, the type of engine found in a petrol or diesel car. Liquid fuel could suddenly propel huge metal vehicles over long distances, and its use expanded rapidly from a very basic domestic context to a global one. At the time, these were positive developments for society, improving living standards and enabling greater mobility. Little was known about the impacts of emitting invisible greenhouse gases into the atmosphere, and the motor car quickly became a fashionable mode of transport, with numbers skyrocketing from a few million cars produced every year in the first half of the 1900s, to over fifty-five million a year by 2000 – a dramatic rise by any measure.[12]

These cars were hungry for fuel, satiated by petroleum extracted from underground reservoirs. When I first thought about one of these reservoirs, I imagined an underground cavern filled with a black liquid, but this is not a realistic image. The substance is actually stored in the tiny micropores of rocks, in a similar way to how a sponge holds water. The oil extracted is often referred to as 'crude oil' and is made up of many different hydrocarbons; as the word suggests, these are chemical compounds composed of hydrogen and carbon. You can visualise these tiny molecules as carbon atoms holding hands, joined together in a chain, with the hydrogen atoms clinging on around the outside. Crude oil is a cocktail of hydrocarbons, varying by the length and shape of the carbon chain and surrounding hydrogen atoms. Hydrocarbons with a short carbon chain weigh less, and hence have different properties compared to heavier molecules with a longer carbon chain. Natural gases have very short carbon chains, so they exist as gas at room temperature, whereas the thick, sticky oil used by ancient Mesopotamians (and Noah to waterproof his boat) would have been made up of long carbon chains.

By separating the crude oil into its various types of hydrocarbons, the best use can be made of the properties of each. But before hydrocarbons can become useful, they have to be found and cleaned up. In the case of natural gas, 'cleaning up' means removing any liquids and other unwanted substances. For oil, things are a bit more complicated. Crude oil is sent to a refinery, where the distillation process takes place to separate it into its constituent parts, separating fluids based on their boiling temperature. Lighter hydrocarbons, with fewer carbon atoms, have a lower boiling point than heavier ones, so lower temperatures are needed to turn them into a gas. If a concoction of crude oil is heated, the smaller hydrocarbons will boil off first. Further along the process, the boiled-off gases are cooled and forced to condense into a tray. This is repeated with the slightly heavier hydrocarbons; they boil off at a slightly higher temperature, and condense into a different tray, and so on until only the heaviest hydrocarbons with the longest carbon chains remain in the mixture.

These different liquids, or 'fractions' as they are known, then get distributed for different uses. Heavy oils are often used to make asphalt – this sticky, black semi-solid material has applications in building roads, its sticky qualities allowing it to glue together the small pieces of stone that make up the road. And, just as the people of ancient Mesopotamia did, it can also be painted onto surfaces, such as the flat roof of a house, to waterproof them. Heavy oils can also be used as fuel for ships, or to generate electricity in power stations. As the carbon chain gets shorter, lighter fractions like kerosene are used as jet fuel for planes. Lighter oils still, like gasoline and naphtha, are used for car fuel or are processed by the chemicals industry into products like plastics. Since these lighter fractions are more useful to us as a society, nowadays much of the heavy oil is broken down into shorter hydrocarbon chains – an energy-intensive process that adds to the carbon footprint of refineries.

But before getting to the stage where we can turn the oil into useful things, the first challenge is to find oil and gas, physically get to it, and extract it.

*

Deep, deep down, in the ground below your feet, oil and gas reserves hide in rock formations. Geologists and geophysicists can map the inside of the Earth with a variety of scientific tricks in order to find these reserves, like a high-stakes treasure hunt. Techniques like seismic reflections – firing seismic waves into the earth and interpreting the reflections that return – are used to create a 3D internal picture, which then allows us to predict the location of oil and gas reserves. But the predictions are just predictions, and the only way to confirm the presence of the substance is to do exploration drilling into the earth, hoping that oil and gas will flow out like a fresh spring in a desert.

This might be the last place you'd expect to find glittering jewels, but diamonds, as the hardest known naturally occurring material on Earth, are critical for drilling. Diamond suppliers provide the (often synthetic) stones for the drill bits used to cut through the layers of rock. If exploration drilling is successful – in other words, if enough oil or gas is found – the next step is to figure out how to actually get the oil or gas out of there. Early on in my career, this was one of my jobs – I was part of a team tasked with figuring out how to get to a pocket of gas that was trapped in rock about three kilometres below the seabed, fifty kilometres off the east coast of England, in an area of the sea with a water depth of fifty-four metres.

While I found my job fascinating and unique, there was always a quiet voice in my head challenging me to think about the morality of it all – the wars petroleum has fuelled, the environmental impacts, and the growing concerns around carbon dioxide emissions and climate change. But I reasoned with the voice for several years, justifying my work with the argument that the world relies on oil and gas. I used it all the time to travel and to stay warm, so I could definitely see and feel its benefits. Someone had to source it, so why not me?

I immersed myself in the technical detail and the company of my knowledgeable colleagues. I learned about the gas journey – from deep underground, under oceans, into pipelines that carry the fluid to a platform, and eventually into other pipes that snake their way back to land along the seabed, and then to an oil or gas processing facility on land and beyond.

Every detail had to be designed and sized based on the properties of the particular oil or gas field – a bit like designing a bespoke suit. What shape and size is the person? What will the weather be like when they wear the suit? How long does the outfit need to last?

Despite the background guilt I felt, my years in oil and gas did train me well for a future career in more sustainable energy production. Many of the skills are transferable, from general technical project management, to knowing the ins and outs of operating a platform in the North Sea – it would be useful information for running an offshore wind farm, for example, if I ever work in that sector. The same applies to other fossil fuel jobs – just as Lynemouth coal-fired power station switched to biomass, so too did the people who worked there. Alongside the useful skills I got from oil and gas, I also experienced living in one of the most unique environments on Earth.

My first trip offshore felt like visiting another planet. I checked in for the flight at a heliport just outside a small seaside town in Norfolk. I was handed a flight survival suit to put on, and eventually managed to get the appropriate parts of my body into the correct locations of the suit, zipping it up to secure myself in. I sat nervously in a flimsy plastic chair alongside my fellow passengers to watch the pre-flight safety video.

The closest I'd got to anything akin to a helicopter before was during the survival training in a swimming pool – I'd never actually been in the air. I filed out of the waiting area, alongside the twelve or so other passengers, and onto the aircraft. With my ear defenders on, the engine started, and the rotors started to woosh around above me, creating a huge roar. I watched the buildings below get smaller as we headed east. The journey was only forty-five minutes, but it took me to a different world.

My new home offshore was a small rectangular metal structure about eighty kilometres off the coast of the UK. It took me about 130 steps to get from one end to the other. As I stepped off the helicopter onto the platform's green helipad, I was greeted by large white capital letters painted on the helipad telling me I was on BABBAGE. The name of the gas field, which I had seen on paper and heard in the office many times, would have been chosen by the geologists or drilling team who discovered it. I was surrounded by endless water, with no land in sight. The metal platform had four floors, with the helideck perched on top. Each of those floors was made from blocks of grating, which meant if you looked down, you could see through the floors, all the way to the sea. I felt strangely secure atop this impressive structure, though I can't say I recommend it for those who are afraid of heights.

I spent the next four days living with nine men on this postage stamp in the middle of the sea. It took a full and slow day for them to adjust to my presence, but they realised I was a normal human being, and accepted me as a colleague. Being a female engineer is uncommon in the places I have worked, so a woman turning up on the platform was a relatively rare occurrence. I bombarded them with questions about the platform, how the equipment worked, how controls systems worked, and how they maintained and operated it all. In return, they indulged me with detailed answers.

I was also curious about the lifestyle for oil and gas platform workers. These men spent two weeks on the platform, followed by two weeks back on land, rotating their time at home and offshore month after month, year after year. They often missed their children's birthdays, family Christmases, and other life events that were happening on land, which worked well for some, but less so for others. One of the men was on his sixth marriage. As well as being difficult in terms of home life, the two weeks offshore are full-on. It is a twelve-hour working day, and there is not a lot to do in the evenings, aside from watching films, playing Monopoly, and enjoying whatever the chef has prepared for dinner before heading to sleep in the boat-cabin-like accommodation. By my last day living in this intense situation, I had built a bond with the group, and my later trips to the platform, once every few months, felt to me like visiting family.

Many people mistakenly use the terms 'rig' and 'platform' interchangeably. The word 'rig' refers to a drilling rig – a special structure that drills wells. They come in different shapes and sizes, but the most common in the offshore environment is a jack-up rig. This structure floats along the water to its desired drilling location. Once it arrives, huge 'legs' at each corner of the rig descend into the water like a three-legged, overgrown spider, until they hit the seabed, elevating the platform and equipment above the water level and providing a flat, stable surface to drill from. The mechanics of it is much like a jack that you use to lift a car when changing a tyre. Once a well has been drilled, it is temporarily plugged, like a cork in a bottle, to stop any oil or gas escaping.

A platform is then installed in place, connected to the well, and the plugs are removed by drilling through them to allow the gas to flow and get piped back to shore. When a drilling rig has finished its work, the legs return to their original position, and the rig floats along to its next drilling job, carrying with it the people who work on it. At the moment, around the world, there are about 1,000 drilling rigs, most commonly found in the North Sea and the Gulf of Mexico.[13]

Platforms, like the one I stayed on, permanently sit over the top of oil or gas wells and act as the receiver of the substance, carrying out some preliminary processing. On the platform I visited, it was a relatively simple operation. Three gas wells produced gas, mixed up with small amounts of liquid, like water or oil. That liquid was separated out from the gas, which was then sent in a pipeline, running along the seabed, to an onshore gas-processing terminal.

The gas wells I was working on had been hydraulically fractured, or fracked. Fracking is the somewhat controversial process of injecting high-pressure freshwater to fracture the rock formations containing the gas, followed by an injection of sand-like particles to hold the fractures open. Normally, drilling a hole into an oil- or gas-containing rock formation creates an easy path by which the substance, which is under high pressure, can escape. It's a bit like shaking up a can of a fizzy drink then opening it; the liquid suddenly has a route out, to release the pressure. But in some cases, the natural pores in the rock are too small to let the oil or gas escape. To coax it out, the rock is hydraulically fractured to increase the size of the pores and let the gas flow out. Very simply, this means breaking (fracturing) the rock by using the force from high-pressure water (hydraulics).

Most people have come across contentious cases of fracking done on land, but its offshore cousin gets little attention by comparison. As my time in the oil and gas industry ticked on, I grappled with my own conflicting views. On the one hand, fracking allows humans to access hard-to-reach oil and gas resources, and has contributed significantly to the US's energy security, enabling the country to rely more on its own resources than imports. But fracking comes with a high environmental cost. Many environmentalists highlight the large amounts of fresh water needed for fracking and the toxicity of the fracking fluids as a major concern, alongside the earth tremors that can result from the operation.

For me, climate change is the strongest case against fracking for oil and gas. Some argue that access to abundant natural gas could act as an incentive to switch from coal-fired to gas-fired electricity generation, which has the potential to halve carbon dioxide emissions per unit of electricity produced. While this might seem like a net positive, it is nowhere near drastic enough to combat the reality of climate change. Hydrocarbons are ultimately better off being left in the ground.

Shortly after visiting the Babbage platform, with my ethical concerns pushed deep down, I joined the hydraulic fracking team. There were plans to bring in a drilling rig to drill two more wells in the field, to increase gas production, and these wells also had to be hydraulically fractured. Some pretty serious equipment is needed for fracking, which is taken to the site by boat. Similar to a drilling rig, these boats are operated by specialist companies, who travel from location to location around the world to carry out hydraulic fracturing jobs. I was aboard the Big Orange XVIII, which didn't quite live up to its name; it was about eighty metres long (so not that big) and it was blue. The plan was for the Big Orange to get as close to the platform as possible, slotting in almost underneath it, and then we'd hook up a hose from the boat to the platform. From there, the crew on the platform would attach the hose into the top of the well, and my boat crew would start pumping fresh water and sand into the well until we detected the rock fracturing. The hose was huge, wound tightly around a reel several times my height. Getting it from the boat to the platform was a complicated operation in itself, carefully planned. The hose was cradled into a lifting sling, and the platform's crane operator deftly lowered the crane arm down to the boat, allowing the boat crew to hook the lifting sling onto the end of the crane arm.

But before the boat could do its fracking job, two other things had to be done: putting in a plug above the previous set of fractures already in place, then puncturing holes in the metal tube lining the well. These are two simple steps on paper, but the jobs are complicated and risky, with impeccable co-ordination needed between the platform and boat crews. Since all of this was taking place deep underground, away from the eyes of the engineers, there was a lot of potential for things to go wrong. As small issues unfolded, the two-day job turned into two weeks of mostly waiting for the platform crew to complete their jobs.

So we waited. The boat had limited internet access and entertainment options. For someone as impatient as me, it was tough to be stuck on a small boat in the middle of the sea for much longer than anticipated. I eventually came to terms with it and got into a routine of learning hydraulic fracturing calculations, writing daily technical reports, binge-watching

Boardwalk Empire, and braving the tiny gym aboard. Running on a treadmill while simultaneously being gently rocked by the water was a strange, somehow soothing feeling. When the kitchen ran out of milk, we had to get creative with what to put on our breakfast cereal; I opted for mango juice, while a colleague went for ice cream (I wouldn't recommend either). As supplies dwindled, the captain decided it was time to take the boat to land to restock. I milled around on the deck, waiting to catch a glimpse of land; the sea was calm, the sky was blue, and the sun was setting over the horizon as we sailed past an offshore wind farm. The turbine blades turned silently and slowly, looking down on me accusingly and filling me with a sense of guilt.

Finally, in the dead of night, our boat team got the go-ahead to hook up to the platform, and our job in the long chain of events kicked off. We worked through the night, and I remember admiring how experienced the team was, and how well they worked together. Max, a Dutch engineer with dark hair flopping over his face, was in charge. I watched as he calmly studied the pressure climbing on a graph on his monitor as water was pumped into the well. As soon as he detected the rock fracturing from the squiggles on the screen, he issued the command to stop pumping. A slurry of gel and sand had been preprepared and tested by the chemist in the tiny boat lab. Max issued another command to start pumping the slurry into the well, using a fine balance of his past experience and our calculations to make the decision on when to stop pumping.

The gas could then finally escape from the rocks in which it had been confined for millions of years. From there, it travelled up to the platform, and back to land, where it went on to be used to power normal, daily lives. Perhaps it heated a home, or cooked a meal, or generated electricity so a switch could be flicked to turn on an appliance, the end user oblivious to its complicated journey and the amount of effort that went into getting that supply of energy to them.

*

It may seem extreme to go to these lengths just to get fossil fuels out of the ground, but these energy-dense substances have a vast array of uses, and we have all become painfully reliant on them. Even with the best of intentions, it's virtually impossible to get through a day of modern life without benefiting, in some way, from fossil fuels. They are everywhere. You are probably wearing or using something that has oil-derived materials in it right now. Synthetic materials, lubricants and fertilisers normally contain oil-derived substances. Plastics, used for food storage, have helped to cut food waste. Until we all decide to give up these products, or find alternative ways of manufacturing them, the future of humans is deeply intertwined with oil and gas.

One of the largest oil users is transport, so if we want to wean ourselves off fossil fuels, we need alternatives to get us around the world with the same ease we are used to. Fossil fuels are used to generate electricity, which could be replaced with renewable generation – but electricity only meets about a fifth of the world's energy needs, as it is not yet used directly for transport and other things like powering industrial processes.

My own life will always be linked to oil, the substance that has determined the fate of the Iraqi nation and its inhabitants. I find it fascinating to look at how and why different nations have dealt with their resources. Norway, another oil-rich country, set up a national wealth fund for its citizens so that profits from drilling for oil benefit everyone, while Iraq, on the other end of the scale, has been plagued by corruption for as long as I can remember. I often wonder what my life would have been like if Iraq did not have oil.

In a strange twist, Farouk al-Kasim, an Iraqi petroleum geologist who moved to Norway in the 1960s, is credited with playing an important role in setting up Norway's forward-looking oil and gas policies. It was never his intention to play this part; he and his Norwegian wife moved to Norway to seek better medical care for their son, who was born with cerebral palsy. Al-Kasim was quickly hired by Norway's Ministry of Industry, at the very beginning of Norway's oil discoveries, and he was instrumental in the government's decision to set up a national oil company, and the careful planning that took place to make sure the oil industry benefited Norwegian citizens.[14] Al-Kasim and the Norwegians showed that there is always a better choice that can be made; there could have been a better story for the rest of the world.

It is now, of course, very well known that the use of fossil fuels releases greenhouse gases into the atmosphere, leading to harmful climate change. Fossil fuels are carbon-based, so when they are combusted, they react with the oxygen in the air. Each carbon atom clings to two oxygen atoms in the air, forming carbon dioxide or CO_2, a greenhouse gas. Methane, which is a major component of natural gas, is about thirty times as potent as carbon dioxide as a heat-trapping gas when viewed over a hundred-year period. Over a shorter timescale, it has an even higher global warming potential, so any methane leaking from oil and gas production operations has an immediate harmful impact on the environment. Looking at a graph of these emissions into the atmosphere over time, you do not need a science degree to spot the enormous rise that coincides with the onset of the Industrial Revolution and the increase in the use of coal, followed by oil and gas.

All of the coal, oil and gas infrastructure will come to an end eventually, and ultimately, the world will run out of fossil fuels. It might take fifty years; it might take 200. But at some point, this resource will end. Oil and gas are finite resources because they take millions of years to form. They cannot regenerate as quickly as they are currently being used up. But it is up to us how soon, and in what ways, we turn away from fossil fuels. As a society, we now have a conscious decision to make between continuing to use fossil fuels, with the knowledge that they harm the environment, or leaving the reserves in their place and finding alternative, renewable sources of energy. Perhaps it is time to say thank you to these dead creatures, who have fuelled our society for generations, and allow them to rest in peace as we continue our quest for better, cleaner energy.

Chapter 2

Nuclear

WHEN I HEAR the word 'nuclear', my brain immediately conjures up images of mushroom clouds, destruction and death. I have to actively reason with myself to get rid of these images, and I think this is the case for many others. Having grown up around the idea of oil and gas, I never thought of them as dangerous, despite both being highly flammable and hazardous. And later, working in oil and gas, I witnessed first-hand the strict safety rules in place. I have never worked in the nuclear energy sector, and perhaps it is this lack of familiarity that evokes the strong emotional reaction. But I think even more than this, nuclear is an energy resource that is poorly painted in history and modern media, despite the positive contributions it has made and the role it can play in reducing carbon dioxide emissions.

The more I learn about it, and meet people who work in the industry, the more this gut reaction fades and softens. Understanding the reason behind the emotions helps to untangle them into something more rational. Nuclear energy was developed backwards – firstly as a weapon with the potential to wipe out huge numbers of people. Power generation came later, a convenient side effect with the advantage over oil and gas of no carbon dioxide emissions, though not without its own, different problems. Given this history, it is no wonder many people feel an inherent fear attached to nuclear power – but we have an opportunity to deal with the fear and make use of a clean energy technology.

The science of atomic radiation began to unfold from the 1800s onwards, but the practicalities of nuclear power truly ramped up during the Second World War as a military imperative. The fear was that Germany was developing nuclear weapons, so the US government threw everything it had at the same goal. But the Nazis' own anti-Semitism ironically hampered their efforts, driving out many of the top scientists in Germany and Nazi-occupied areas, many of whom fled to America.

World-leading German institutes of higher education began to divide physics into two nonsensical and racist categories: 'Jewish physics' and 'Aryan physics'. 'Jewish physics' included Albert Einstein's theories of relativity. Werner Heisenberg, a German Christian, was harassed and labelled a 'White Jew' for his work in quantum mechanics, which was also seen as 'Jewish physics'. Even more bizarrely, Heisenberg's mother got involved. She contacted Heinrich Himmler's mother, whom she knew through family connections, and asked her to tell her son, a leading member of the Nazi party, to step in and give Werner a break. Himmler eventually forbade any further attacks on the physicist. Though Heisenberg got somewhat let off the hook for daring to study specific areas of physics, these political interruptions set back Germany's progress, both in terms of physics as an overall field and more specifically in the development of a deadly weapon, giving the US an advantage.[1,2,3]

The US, being far away from the epicentre of the war, gained another advantage over Germany; it became a safe haven for rejected geniuses and the centre of everything nuclear. The superheroes of physics flooded in: Albert Einstein moved over in 1932, and Johnny von Neumann, thought to be Einstein's intellectual equal from Hungary, had arrived a couple of years earlier. Hans Bethe, a theoretical physicist raised as a Christian but with a Jewish mother, went over in 1935, as did Edward Teller, a Hungarian Jew. Not all who escaped to the US were Jewish; George Kistiakowsky went there to escape the Communist Revolution in Russia. Enrico Fermi, a Roman Catholic Italian and co-inventor of the nuclear reactor, fled Fascist Italy in 1938 because his wife was Jewish. Leo Szilard, an eccentric Hungarian Jew known for developing the idea of a nuclear chain reaction, moved to the US the same year. In 1943, the Danish father of quantum mechanics, Niels Bohr, was the last of the greats to cross to the US.[4]

Germany's loss was to the benefit of America. Many of these physicists worked on the Manhattan Project, the research and development programme that led to the production of the first nuclear weapons, running from 1942 to 1946. As part of the project, Enrico Fermi carried out a secret nuclear reactor experiment in some abandoned squash courts at the University of Chicago. As it transpired years later, the secret wasn't so secret after all, as the Soviets had accurate information about it. They knew the location and subject of the work, but got one thing wrong: they had translated 'abandoned squash courts' to a perplexing 'unused pumpkin patch'. Fermi's experiment proved the concept of a nuclear reactor on 2 December 1942. The American physicist Leona Woods, the youngest person and only woman in the group, crammed into the viewing gallery of the squash courts to witness the experiment, and was also the only one who had the sense to take notes, leaving an important minute-by-minute account of this historic experiment in her notebook.[5]

Over the next few years, two atomic bombs were designed at frantic speed: Fat Man and Little Boy. Kyoto in Japan was first on the list of cities to bomb, but Henry Stimson, US Secretary of War at the time, had visited Kyoto – some historians think it was for his honeymoon – and liked it. He insisted on the selection of a different target. For me, this is a strong demonstration of the benefits of people travelling and getting to know other cultures. Perhaps no bombs would have been dropped if Stimson had spent more time with other Japanese communities. Unfortunately, Stimson apparently did not like the next location on the list quite as much: Hiroshima, a port city close to the south end of Honshu, Japan's main island.[6] On 6 August 1945, the innocent-sounding Little Boy dropped there, causing a temperature burst of more than a million degrees Celsius and vaporising everything in the way. Anything that was spared the initial burst caught fire as air rushed in to replace the space the fireball took up in the atmosphere. Initially, the Japanese government was confused – why was no one answering the phones in Hiroshima? Little did they know the phone system had been completely destroyed. Electricity demand for the city of 350,000 plummeted to zero. Around 80,000 people died instantly, and that same number again died later from the radiation and aftermath.[7] This was not the end; another bomb followed three days later.

On 9 August 1945, the jovially named Fat Man was dropped on Nagasaki, where more than 70,000 people instantly died.[8] Both bombs wrought unspeakable destruction – so much so that the design specification and drawings for Little Boy were burned on a bonfire in the US after the bomb was dropped on Hiroshima to stop others building more.[9]

But rather than putting a complete stop to nuclear weapons development, the fear, death and destruction spurred it on into the Cold War. In 1955, twenty nuclear weapons were tested above ground. The number went up every year, reaching 140 weapons set off in the atmosphere in 1961, mostly by the US. Sometimes multiple bombs would be tested on the same day, leading to a cocktail of dangerous radiation in the atmosphere, which built up in the materials it made contact with. It became impossible to hide the radioactive fallout from the public; children were even told not to eat snow for fear of the radiation build-up,[10] cementing the image of nuclear power as a danger to society. Using it as a clean energy source – as a way to help humanity, rather than destroy it – was the furthest thing from people's minds.

*

In 1963 there came the first glimmer of hope: the US, UK and Soviet Union signed the Limited Test Ban Treaty, making nuclear weapons testing in the atmosphere, outer space and underwater illegal. As of 2022, there were around 10,000 nuclear warheads in the world – more than enough to destroy humanity as we know it, but a dramatic decline from the 60,000 or so warheads around at the peak of the Cold War.[11]

Talk of using nuclear energy to generate electricity had been simmering along behind the scenes for longer than we might imagine. Following the Second World War, the Americans, Brits, Canadians and Soviets all poured efforts into power generation from nuclear. The first civilian nuclear power station fed electricity to the grid on 26 June 1954, in Obninsk, Russia, less than a decade after Little Boy and Fat Man wreaked havoc on Japan.[12]

To understand how nuclear power generation works, we have to dip our toes into the strange world of particle physics, and the general idea that everything is made up of tiny particles. The earliest thoughts on this subject were recorded in India in around 550 BCE, when the schools of Indian philosophy of Nyaya and Vaisheshika described how particles of matter combine to make more complex materials. The Greeks ran with this same idea about a century later, believing that matter was made up of invisibly small pieces, which could not be broken down any more. These were named *atomos*, meaning 'uncuttable' and it is where the word we use today, atom, comes from.[13]

Perhaps unsurprisingly, the ancient Greeks didn't have it completely right. It turned out that these atoms *are* cuttable in a way, as each is made up of even smaller particles – a central nucleus of positively charged protons and neutral neutrons, with tiny negatively charged electrons floating around the central nucleus.

Just as the Greeks didn't have it completely right, as scientific understanding progresses, this picture will improve. In a similar way to how the Earth and other planets in our solar system are held in place and orbit the sun thanks to the forces of gravity, there are forces at play that hold protons, neutrons and electrons together at the very small scale, and, on a slightly larger (but still tiny) scale, there are forces that hold atoms together. These atoms make up the copper in cables used to move electricity, lithium in batteries for storing energy, and everything else around us.

Breaking the forces holding subatomic nuclear particles together requires a huge amount of energy, but doing so can release an even larger amount of energy. This may seem counterintuitive, but it begins to make sense if we start to consider Einstein's groundbreaking discovery: that energy and mass are interchangeable. The laws of thermodynamics say that energy and mass cannot be created or destroyed, but mass can change itself to energy, and vice versa.

In 1903, scientists Ernest Rutherford and Frederick Soddy showed this theory in practical experiments. Based on their experimental measurements, they calculated that millions of times more energy would be released from breaking down the atoms of one gram of the radioactive element radium, compared to burning a gram of fuel.[14] Chemical reactions, like burning petrol, play with the electrons floating around the outside of a nucleus, whereas nuclear reactions get into the core of the atom. The forces that glue the insides of the nucleus of an atom together are much stronger than those on the outside; it takes more effort to break them, but in turn it releases more energy than simple combustion.

Delving into Einstein's famous equation, the loss of mass from burning fossil fuels – a chemical reaction involving the outer electrons around the nucleus – is so small it can barely be measured. In practice, this means that the energy you can harvest from chemical reactions is relatively small. But rearrange a nucleus by splitting it (fission) or sticking protons back together (fusion), and the mass lost becomes much more noticeable. Since mass is being converted to energy, more mass lost means more energy is released. In theory, the mass of one raisin converted into pure energy could power a major city like London for a day – but this is, unfortunately, purely theoretical; we do not yet have a method for converting matter, in this case a raisin, directly into energy.

Putting theoretical raisins aside and moving back to the real world, radioactive elements like uranium, mined from the ground mainly in Kazakhstan, Australia and Canada, are a more realistic option.[15] Nuclear power stations rely on the process of fission, splitting atoms, to release energy. Radioactive elements, like uranium, are unstable, and their nuclei have a tendency to break down into another, more stable type of nucleus. For some materials, this takes a fraction of a second, and for others it takes billions of years. The liberation of these radioactive substances reduces over time, with the 'decay' releasing particles and radiation – in a nuclear power station, the nuclear reactor accelerates this process and contains this reaction inside the structure of the reactor.[16]

Out in the wild, we are all regularly exposed to radiation from the sun and naturally occurring sources of radiation in the environment. Living in a building made from stone, brick or concrete, or taking a flight, exposes you to radiation. Radiation exposure also comes from some medical procedures, like getting X-rays. This is very normal, and it only becomes an issue if the dose is very high – high exposure can kill cells in the body, or cause mutations to DNA, which could turn cancerous. But the key to keeping us safe from nuclear reactors is containing the radiation within specially designed areas, and there are strict limits on how much radiation is allowed to be released.

During fission, more mass is converted to energy than in chemical reactions like burning petrol or wood. This creates an advantage for nuclear power, because smaller amounts of radioactive material are needed to generate electricity when compared to fossil fuels. There is two to three million times more energy contained in a kilogram of uranium than in a kilogram of a fossil fuel.

Once it is mined and processed, bright yellow uranium oxide powder is eventually turned into small fuel pellets and heated to make a hard ceramic material in special fuel fabrication factories. The circular pellets are stacked together inside a sealed metal tube, called a fuel rod, like Trebor mints packed into a paper tube. A few hundred fuel rods are then bundled together to make a fuel assembly – like many packs of Trebor mints stacked in a box. Each fuel assembly is several metres long, and each nuclear reactor core in a nuclear power station is made up of a few hundred of these fuel assemblies.[17, 18] As nuclear power station designs evolve to become more efficient over the coming years, the details of the fuels will also adapt.

Most uranium atoms have a total of 238 protons and neutrons glued together in their nucleus, but about one in every 140 uranium atoms are missing three neutrons. The lack of neutrons makes these uranium atoms, known as uranium-235, less stable than their heavier sisters; firing a neutron at this nucleus destabilises it, causing it to split into some lighter atoms and some stray neutrons. During this fission reaction, right as the heart of the atom splits apart, some of the mass is transformed into energy – this is the key to nuclear power.

The stray neutrons shoot off; if they hit more nuclei, the fission process is repeated over and over – a chain reaction that can release enormous and destructive amounts of energy. This is the property that makes nuclear fission reactions good for bombs, and why my brain automatically associates the word 'nuclear' with destruction. But the fuel pellets for nuclear power stations only contain up to 5%[19] of these unstable uranium-235 atoms (compared to weapons-grade uranium, which has over 90%) and there are a series of safety features to control this process and stop a full-scale meltdown in a nuclear reactor, including the control rods.

A key safety feature of any nuclear reactor, control rods are used to catch some of the neutrons to stop a runaway chain reaction and control the fission process. They can be pushed into the reactor to mop up more neutrons and reduce the reaction rate, or taken out to increase it. Most neutrons released from the fission reaction are too fast to be captured by nuclei, so the reaction can actually fizzle out – quite the opposite of the dramatic chain reaction imagery.

The moderator, another important feature of a nuclear reactor, slows down and scatters the neutrons so they can be captured in the first place, keeping fission going. Most reactors use purified water as the moderator, and so the fuel rods are surrounded by water, which slows down neutrons but also cools down the whole process. Beyond the nuclear physics in the reactor, electricity is generated in a very similar way to any other thermal power station, where heat turns water to steam, which goes on to spin a turbine and a generator.

By the late 1960s, nuclear power generation boomed, with new power stations going up around the world. But even as nuclear became more established in the 1970s and 1980s, national debates passionately raged, arguing for or against 'nuclear' power stations, ignited by a small number of dramatic and terrible accidents. On 16 March 1979, a fictional disaster thriller called *The China Syndrome* was released in US cinemas, starring Jane Fonda, Jack Lemmon and Michael Douglas. The movie told the story of a television reporter and her cameraman unearthing safety cover-ups at a nuclear power plant. The term 'China syndrome' is a hyperbolic description of a nuclear meltdown, where the reactor melts through the Earth all the way to China: a fitting term for the time, as national debate over the use of nuclear energy raged. Upon the movie's release, one executive of a company that made nuclear reactors called the film 'an overall character assassination of an entire industry.'[20]

Twelve days on, *The China Syndrome* was still showing in theatres when a Hollywood-style drama began to play out in real life. In Pennsylvania, part of the Three Mile Island nuclear power station's core melted. The power station experienced a series of errors that almost resulted in a radioactive meltdown, but the disaster was avoided when the operators realised that they needed to get cooling water moving through the core, bringing temperatures back down. More than half the core was destroyed, but the outer protective shell was intact, keeping the radiation contained. No one was killed or injured by the accident, and subsequent health studies did not pick up any increase in exposure to radiation within the vicinity of the power station.[21] This does not diminish the seriousness of the accident, but it was not a major disaster on this occasion – the Hollywood drama remained fiction.

Seven years later, on 28 April 1986, the high radiation alarms went off at Forsmark nuclear power plant in Sweden. Staff immediately started looking for the source of the radiation leak. After a thorough scan, they realised that it was actually coming from 1,100 kilometres away, following an accident that had occurred two days earlier.[22] It was, in fact, the fallout from the worst nuclear power station disaster to date: Chernobyl.

Soviet authorities had been downplaying the situation in Chernobyl as much as possible – until Sweden exposed the scale of the disaster, forcing them to acknowledge it. The official death toll for the disaster sits at fifty, but thousands of people were exposed to non-fatal doses of radiation, causing deaths and severe illness later down the line. A radioactive cloud blanketed Europe, and a clearer picture of the accident started to emerge after the collapse of the Soviet Union in 1991.[23, 24] As well as having a catastrophic impact on people's lives, their health and their environment, Chernobyl has left a deep impression in minds across the globe.

After such a catastrophe, fear ran deep in the public consciousness, and I imagine the world was on tenterhooks, waiting for the next one. But it didn't come for another twenty-five years. During that time, another fifty or so reactors were built around the world, and the dread of a Chernobyl repeat subsided.[25, 26] But disaster struck again in 2011 in Fukushima, Japan. On 11 March, an earthquake and tsunami struck the main island of Japan, causing destruction and thousands of deaths. It was the strongest earthquake Japan had experienced in 1,000 years, and came with devastating knock-on effects, including the flooding of the Fukushima Daiichi nuclear power plant, at the time one of the largest nuclear power stations in the world.

The power supply failed and operators lost control of the plant. Three of the reactors overheated to the point of nuclear meltdown. The high temperatures resulted in the production of hydrogen. Mixed with oxygen, this blend exploded, damaging the building housing the reactor. While the reactor core did melt, it was contained within the safety structure; however, some radioactive materials were released with the hydrogen explosion.[27] The accident remains one of the worst nuclear disasters in history, but it is difficult to untangle the harm done by the nuclear accident from the effects of the natural disasters that led to it. The incident changed the face of the nuclear power industry once again, bringing the risks back into the forefront of our minds and causing some governments, like Germany's, to make the decision to close down their nuclear power stations – a decision they may come to regret from a carbon emissions perspective.

When nuclear power stations go wrong, it can be catastrophic. But despite the deep and long-lasting impressions left in our collective memory by the tragedies, the probability of these events occurring is extremely low. When I think about Fukushima or Chernobyl, I easily spiral into a place of worry and fear, so I consciously try to put it into perspective. The business of energy generation can be harmful to human life in multiple ways. For example, air pollution leads to long-term health issues and premature deaths all over the world. On a list of death rates per unit of electricity produced, coal and oil come out on top – about 1,000 times higher than nuclear power, which has numbers comparable to deaths from solar and wind power generation.[28] Of course, this is a complicated assessment to make, and depends on how deaths are attributed and where the lines are drawn. Lower numbers of deaths do not in any way make the loss easier, but it is important to consider the harm done by different types of energy generation side by side, because the fact of the matter is that we need energy to survive, and this puts the safety of nuclear power into perspective.

Given the fear and gloom surrounding nuclear power, it is easy to dismiss it as an option for power generation, despite the benefits. It is also easy to forget what an incredible leap of science and engineering it was. Perhaps if we had a chance to go back and change history, events may have unfolded in a different way, and nuclear power could have been an ideal solution to the world's energy problems. It converts matter directly to energy, without burning, and uses small amounts of an available fuel, without releasing greenhouse gases.

The excitement of nuclear physics discoveries and the breakneck speed of development during the Manhattan project are now long gone. The technology has improved since then, and there are strict standards on allowable radiation doses, and a general ethos around the importance of safety. As a result, nuclear power is a slow and, some might say, boring industry. But that is what I want from this hazardous source of energy. It's what I want from any high-risk industry out there.

Despite the lingering history, threat of weaponisation and devastating accidents like Chernobyl and Fukushima, nuclear power has been generally – if not widely – accepted. Today, there are around 440 nuclear power stations in just thirty-two countries around the world, pouring about 10% of the world's electricity needs into the system.[29] But one issue that remains a source of concern is the nuclear waste that is created in the process. Spent nuclear fuel remains dangerously radioactive long after it has been used up, so it is initially kept underwater. Like some of the scenes of Homer Simpson doing his job in *The Simpsons*, the fuel is moved by remote controlled devices, operated from behind a radiation shield, into a concrete and steel swimming pool to shield workers from the radioactivity, where it sits for several years. When it has cooled down, it is moved into another pool, or into steel cylinders, closed off, and kept in there for another forty years.

The radioactivity persists for thousands of years, so careful disposal is imperative to protect people and the environment, but as it stands, the vast majority of spent fuel has nowhere to go. It remains the ethical and legal responsibility of the country that produced it – it will not be sent across borders for disposal.[30] After all, who would want to live near radioactive nuclear waste?

*

Finland's internationally renowned opera singer Mika Kares stands in a grotto-like cave, bathed in an eerie blue light. His smart outfit – black suit, white shirt and white bow tie – is somewhat ruined by the fluorescent high-vis vest under his black jacket, and the green hard hat and safety goggles. His deep voice reverberates in the chamber, and the notes from his accompanying keyboard player dance off the walls of the grotto. People watching his performance as part of the Bel Canto festival on television in Finland were puzzled – where is Kares's mysterious grotto stage? Perhaps a few were able to guess that Kares was, in fact, inside Finland's permanent nuclear waste disposal facility, Onkalo.

Pasi Tuohimaa, communications manager at Posiva, the company that is building and operating the facility, laughed as he indulged me with the details of this strange scene.[31] When I asked Pasi about how he ended up working for Onkalo, he explained that he has always been pro-nuclear, so after a thirty-year career of journalism in Finnish media, he was excited to move over to this job. I found a real sense of pragmatism and practicality conveyed by Pasi, which appealed to the engineer in me. At the moment, he told me, most nuclear waste across the world is in temporary storage at ground level, but Pasi strongly believes that leaving it there for the next generation to deal with is not responsible or fair – we created this waste, and we should deal with it to the best of our abilities. I found myself nodding along.

Finland has spent forty years trying to figure out a responsible way of dealing with nuclear waste. They considered shooting the waste into space, or burying it in polar glaciers. They thought about recycling the waste fuel, which sounds appealing but is technically challenging, and expensive and does not get rid of all the radioactive waste. In the end, they settled on deep geological disposal – storing the waste in underground stable rock formations. 'With all the knowledge that exists in the world, it is the best solution,' Pasi summarised philosophically.

A few locations in Finland were investigated as potential disposal sites. Geological surveys, socioeconomic impact assessments, the logistics of moving spent fuel to the site, and the existing infrastructure were all taken into consideration. The final rock formation that was chosen, in western Finland, is two billion years old. This is beyond the timescales of human history. During those two billion years, dinosaurs and all sorts of creatures have come and gone, while the rock formation has stayed the same.

Nuclear waste goes through a series of stages before it is ready to be 'disposed of'. At the time of final disposal, only a thousandth of the fuel's original radiation remains, and it takes about 250,000 years for radioactivity to drop down to the level of a uranium ore deposit – a short period compared to the age of the rock formation.[32] During the forty years of research, scientists and engineers took account of absolutely everything they could think of – future ice ages, earthquakes, sea level rises. In every scenario, the rock formation stays the same. 'What could be a safer place on earth?' Pasi asked me.

The idea of burying something in rock for up to a million years, essentially forever, might seem ludicrous at first. But equally, this is the best idea we have for the time being, and it is better than leaving the nuclear waste in temporary storage, where it is truly exposed to earthquakes, sea level rises and the next ice age. A literal and metaphorical back door is left open; the spent nuclear fuel can be brought back out in the next hundred years in case a better solution comes to light.

Onkalo itself consists of a series of tunnels: a spiral-shaped access tunnel, four vertical shafts for people, nuclear waste, and ventilation, as well as the nuclear waste storage tunnels. By 2020, about ten kilometres of different kinds of tunnels had been excavated, with around thirty kilometres more to follow as the site develops. I imagined endless barbed wire fences and menacing-looking warning signs around this facility, to try to warn future generations from tampering with the buried poison. But how could you communicate this message to people 1,000 years from now, who will likely have very different lives and languages? After all, even today, when we find writing we can't understand from ancient times, our instinct is to dig. But Pasi surprised me: 'The concept is that you don't need to communicate it.'

Spent fuel will be put into the tunnels over the course of the next hundred years, he explained. Once the disposal facility is full, it will be permanently closed up and sealed off. It's remarkably difficult to dig through half a kilometre of rock. If an ice age happens, the area will likely eventually turn into forest. There are no valuable materials to be mined – at least, nothing that we consider valuable today – so the idea is that no one will know it is there or have a reason to go there. There are many unknowns with this assumption; after all, perhaps future beings will want to dig up this area for new reasons that we can't even begin to imagine from the twenty-first century. But for the moment, this is the best we can do.

The facility has been built to house the radioactive waste generated by the nearby Olkiluoto nuclear power plants. The fuel assemblies from these power stations, weighing about a quarter of a tonne each, provide energy for the reactors for around four years. Once they are spent, they are removed and placed into the pool and interim storage for forty years, before going off to the encapsulation plant to get packed up for their final resting place. Much like the multiple layers of safety precautions in nuclear power stations, packing the waste relies on the idea of multiple independent layers of protection, so that if one layer fails, the others will still be there to stop the radiation from escaping into the living environment.

The state of the fuel, as ceramic pellets made by compacting and heating the uranium powder, constitutes one layer of protection in itself. The pellets are packed up inside gas-tight metal rods, another layer of protection. The next layer is a canister made of a special type of cast iron surrounded by a copper shell to ward off corrosion. When the fuel is inside the canister, it is filled up with argon – an unreactive gas – and sealed. The canisters are always a metre in diameter, but the length varies between three and five metres, depending on where the spent fuel comes from. The canisters are then placed about 430 metres underground, adding another barrier of the ground itself. That is not all, though – once the canister is in place underground, the surrounding space is filled with bentonite, a clay that fills any empty space between the canister and the rock to protect it even more, absorbing radiation in the unlikely event it escapes its prison. The canisters are stacked next to each other in the tunnels, and when a tunnel is full, the whole thing is closed up with a steel-reinforced concrete plug.

A hundred years from now, in the 2120s, all the tunnels will be full and all the access tunnels will be closed. Any connection with ground level will be sealed off, as if nothing ever happened. The hundred-year lifetime has been calculated back from the expected operating time of the local nuclear power station, Olkiluoto, which will generate electricity for another sixty years, with the spent fuel needing forty years in the pools before it can go into permanent storage. After that, who knows what the world will be like and where the energy supply will come from? The world has changed a lot over the past hundred years; it is likely to change rapidly over the coming century too.

Similar permanent nuclear waste disposal facilities are being planned across the world. The Swedes are following suit, after the government gave permission for construction of their own facility in January 2022, and France is looking into something similar.[33] This isn't a solution every country can adopt, though – it relies on a number of factors, including having a secure, stable and deep bed of rock to carve into. As well as this, a key issue is the acceptance of the local community. Pasi explained to me that the technology for final disposal is not new, but for any government or organisation thinking about putting in a permanent nuclear disposal facility, there is a huge education and trust-building exercise to go through. This takes time, and Finland has been doing it for forty years. 'All the people who live in Eurajoki or around that area have somebody in their family or their friends who have worked with us,' Pasi explained. They get to know the safety culture, and the more they know, the less they fear.

The information does not just get passed on by word of mouth. The nuclear industry on the west coast of Finland made a conscious decision back at the start of their operations, in the 1960s, to be open and transparent. There are bus tours to the site and visitor centres, neighbouring municipalities are invited along to learn about the facility, and there are science camps for children. They even sponsor the local ice hockey club. It is easy to throw money at the local community, but harder to patiently open up and honestly explain what is going on and why. There is another angle to this, too. The Finns know that nuclear is an important part of their energy mix, a part of their lives. Most of the heating in homes here is generated from electricity, rather than gas like in the Netherlands or the UK. It is not the time to be scared of nuclear energy when it is minus twenty degrees Celsius outside.

In its underground store, the radioactivity of the spent fuel reduces over time, and will be down to naturally occurring levels after 250 millennia. 'Compared to the global climate threat, the question of what to do with the used nuclear fuel is a minor question,' Pasi commented. Thinking over what he has told me, it feels like a case of perspective – one solution may not appear perfect on the surface, but how does it look compared to the alternative?

*

The future of nuclear is potentially brighter than its past. Following Finland's example and rolling out the best-available waste disposal solution opens up the opportunity to use some newer nuclear technologies, like advanced and small modular reactors. These new types of reactors are smaller than today's conventional nuclear power generators, and they are designed so that much of the power station can be built in a factory and assembled at the final location of the power station, rather than being constructed onsite. This means that identical small nuclear reactors can be pumped out of a factory and moved to the site, reducing the cost of putting up a nuclear power station. Identical reactors also mean more standardised safety and disposal protocols, but as with anything, there may be disadvantages – some researchers believe small modular reactors will produce more radioactive waste than today's nuclear reactors.[34, 35]

Amongst all this, there is one radically different option. Until now, nuclear power generation has been all about fission – splitting atoms apart. The holy grail of energy generation is the opposite of this: fusion, or essentially jamming nuclei together to release energy. Mimicking the reaction that takes place on our sun, fusion takes tiny hydrogen atoms and fuses them together into larger ones. But the two atoms weigh more individually than the combination of the two. The loss of weight shows itself as energy – it is a direct mass-to-energy conversion, spraying the universe and our planet with energy.[36]

'I often say fusion is the energy of the stars, and scientists are working to harness it here on Earth,' Dr Melanie Windridge, a physicist and the founder of Fusion Energy Insights told me over a video call.[37] Melanie's six-week-old daughter was strapped to her in a baby carrier, quietly sleeping through our conversation.

After studying physics, Melanie travelled the world, wondering what to do with her life. 'I was doing things like hiking in the Himalayas, and seeing glacial retreats. I did a lot of diving, and you see coral bleaching on the reefs,' she recalled, talking about the deep impression left on her by the experience of witnessing the effects of climate change and environmental damage first-hand. This is what sparked her interest in fusion as a potential solution for the world's energy and climate problems.

Fusion sounds simple on paper – smash some nuclei together and job done – but it turns out it's very difficult to start the reaction, and even more difficult to keep it going. The conditions have to be perfect, otherwise the reaction just stops, making fusion energy inherently safe. There is no chance of the kind of destructive runaway reaction associated with an atomic bomb.

The fuel for fusion energy could be plentiful. For a start, only a small amount of fuel releases a large amount of energy. The energy from an amount of fusion fuel the size of a bag of sugar is comparable to that from truckloads of coal. However, getting the reaction going requires a combination of a type of heavy hydrogen called deuterium, which is found in seawater, and lithium (also present in seawater, although we do not yet commonly extract it from here). The lithium is not a fuel itself, but it is needed to make another heavy hydrogen essential for the fusion reaction, which doesn't occur in nature: tritium. Converting lithium to tritium needs neutrons, and for the time being, the only practical source is from a nuclear fission reactor – so fission reactors will be needed for fusion, until another solution is developed.

While lithium exists in seawater and the Earth's crust, there are challenges. According to Melanie, getting it 'could become a problem because there is competition for lithium for other things like batteries', but it is too early to say. On the plus side, at the other end of the process, no waste is produced by the reaction itself. The final by-product is the inert gas used to fill party balloons – helium. While much of this is still unknown because it has not been done at large scale yet, fusion power plants are expected to produce larger quantities of less radioactive wastes, coming mainly from the structures – like the steel in the fusion machines themselves – so careful disposal would still be needed.[38] However, fusion waste will become safe much quicker than fission waste – in tens to hundreds of years rather than thousands. Scientists are working to minimise the radioactivity of future fusion power plants and figure out the answers to many of these issues.

While nuclear fusion has been happening for over four billion years on our sun, it has so far not been possible to replicate the same reaction on Earth, although scientists are getting closer. For nuclei to fuse on Earth, extraordinarily high temperatures are needed – over a hundred million degrees Celsius, which requires a huge input of energy. But the exercise is pointless unless more energy can be generated than is initially put in. The fusion community recently achieved the momentous energy break-even point, where the energy output exceeded the input.[39] While this proves that the science of fusion can occur on Earth, sustaining and containing the hot subatomic soup of particles, the plasma, is a challenge, to say the least. There are different concepts for doing this, but one of the most advanced solutions is the 'tokamak', a Russian acronym which translates to 'toroidal chamber with magnetic coils'. 'Toroidal' is the mathematical way of saying doughnut-shaped, and the magnetic coils provide magnetic fields to trap the plasma inside the doughnut.

'The principle there is that you hold this hot plasma away from the walls of the machine using very strong magnets. You don't let the fuel hit the wall,' Melanie explained to me. The primary purpose of stopping the extremely hot plasma from touching the – relatively speaking – cold walls of the doughnut is to prevent the plasma from cooling down too much and killing off the reaction. The combination of magnetic fields causes the plasma to move around the doughnut in a helical twist, so it doesn't drift outwards and escape to the walls. More than 200 tokamaks of varying sizes have been tested over the years, but so far plasma has typically only lasted for minutes, at most.[40] Minutes might not sound very impressive, but this is a vast improvement from the mere microseconds achieved about sixty years ago. Once we figure out how to tame and maintain the hot plasma, it is the sustained heat that can be used as energy. It could be used directly, as heat for industrial processes, or it could be used to boil water into steam and drive a turbine for electricity generation.

Many issues still need to be worked out before effectively building a sun on Earth becomes reality. But necessity is the mother of invention, as Melanie points out to me, quoting Russian scientist Lev Artsimovich who once said that 'fusion will be ready when society needs it'. That time may be now, and Melanie sees a turning point coming, driven by the fact that the science is now mature, and there is increasing urgency for alternative energy resources, both to combat climate change and for energy security reasons.

Given the magnitude of the climate crisis we are facing, it seems inevitable that nuclear power generation will have a part to play. But, as Melanie emphasises, it is not the answer to all our energy problems; it's only a piece of the puzzle. Before it can be widely accepted and therefore widely used, we have to acknowledge its history and somehow decouple it from war and weaponry. Without a sincere effort at education and open conversation, nuclear power will remain a dark and mysterious force, feared and loathed by many. Communicating the benefits and dangers clearly and honestly, with care and consideration, will allow people, myself included, to accept nuclear power as a useful and carbon-free source of electricity that can serve us for generations to come alongside other renewables.

Chapter 3

Sun

THE SEAT SCORCHED the back of my legs, and I instantly jumped back up after climbing into the car. It had been baking in the sun all day while my family and I cooled off in one of Baghdad's swimming pools. The competition for the few shady parking spots was fierce, and we did not win it that day. My father pulled out a towel to cover the steering wheel, which was too hot to touch for the drive home. I spent the first few years of my life in this city, where temperatures climb as high as fifty degrees Celsius in the sweltering summer months. My memories of this time are faint but still dominated by the sun. I remember palm trees, always playing outside, and small green geckos lounging on shaded walls. I can still picture the tiny dust particles that glittered in beams of sunshine streaming through the window, making it seem as if I could reach up and touch the sunlight.

It was the early 1990s; Saddam Hussein's power had gripped Iraq for over a decade. Coupled with sanctions against the country, every new day brought new challenges for the population – the lack of freedom of speech and fear of persecution, as well as the difficulties of getting basic daily necessities like water and power. As a child, I was mostly blissfully unaware; my parents shouldered the burden and kept me and my siblings sheltered from it all. But some of the more bizarre shortages were obvious, even to me; I knew that bananas and ice cream were hard to come by (making them extra enjoyable, even to this day).

Eventually my parents had enough of the compounding difficulties of life and made the tough decision to get out, leaving their homeland, family and friends behind. From Baghdad, we relocated to another sunny location: Benghazi in Libya. Again, I only remember the sun beating down on me, and beautiful sandy beaches. We must have had winters, and temperatures do drop to around ten degrees Celsius in both Baghdad and Benghazi, but I don't remember ever wearing a jacket or feeling cold. The sun and its warmth filled every corner of my life and memories.

At the time, I didn't fully appreciate the value of the sun as the reason for life on Earth as well as its potential as an energy source – it took many years of work in the energy industry for that realisation to form. Without sunlight, life would not exist on our planet. We are here thanks to our perfect distance from the sun – not too close and not too far. Most of the energy we consume can be traced back, in some way or another, to the same hot sun that made me jump out of my car seat that day in Baghdad.

Fossil fuels like coal, oil and gas first existed as plants and other living organisms. These plants converted energy from the sun into chemical energy by photosynthesis, enabling them to grow and reproduce. Other living organisms, in turn, ate the plants to get their own energy. When humans discovered fossil fuels, we essentially unearthed a huge and ancient store of solar energy. And through much of human history, we have used plants directly for fuel too, by collecting and burning wood. To this day, energy from the sun captured by trees is routinely turned into logs for fireplaces or wood pellets for heat or electricity generation.

Even the wind that spins wind turbines to generate electricity blows because the sun unevenly heats the surface of the Earth, moving air from one place to another. The variable topology of our planet – valleys, mountains, lakes, rivers and oceans – absorbs the sun's radiation unequally, heating up and radiating heat back out, warming the atmosphere unevenly. As warm air rises, colder air moves in underneath to replace it. The constant uneven heating effect gives rise to wind that sweeps across the seas, causing the water to ebb and flow as waves – this movement, too, can be captured by devices like wave energy converters to generate electricity.

There are just a few exceptions; a handful of energy sources that are not directly attributable to our sun. Energy in the rise and fall of the tides is caused by the gravitational forces between the Earth and the moon, with the sun playing only a supporting role. The other exception is nuclear power, which developed as we gained a deeper understanding of nuclear physics and the components of atoms. The end result, heat or electricity, comes from harnessing the energy released when radioactive materials decay. Likewise, geothermal energy – heat stored in the Earth – originates from the formation of the planet and the radioactive decay of materials.

Humans have always found ways of harnessing solar energy stored in forms like fossil fuels or wood, but we can just skip out the go-between sources and use the sun's energy directly. An enormous amount of solar energy hits the Earth directly: around 180,000 terawatts (TW), many times more than the energy we need.[1] In fact, enough solar energy reaches the Earth every hour to meet the energy needs of the global population for an entire year.[2] This is an enormous untapped potential that we only just seem to be waking up to. While this makes it sound as if our energy problems are easy to solve with the sun, of course capturing this energy, and getting it to where it is needed, is easier said than done.

Two of the countries I called home during my childhood, Iraq and Libya, get plenty of sunshine; more than enough, even, in terms of pure energy. Despite this, to the detriment of the environment, they rely almost entirely on fossil fuels to meet the energy needs of the population, and have done so for over a century. And it is not just their own needs being met; Iraq and Libya's fossil fuels have sustained the energy needs of others for almost as long as we've known about them. Their past, present and future could have been entirely different had they harnessed the energy of the sun rather than fossil fuels.

Unlike recent history, many ancient civilisations understood the value of the energy coming from the sun and made good use of it. The ancient Chinese – stretching back to the Bronze Age – are credited with inventing an ingenious device to use the sun's energy, the *yang-sui*, or 'burning mirror'. The curved bronze disc was used to concentrate the sun's rays so intensely that directing the mirror at an object would make it burst into flames, equivalent to matches or lighters today.

In ancient China, people also recognised that homes and palaces could be cleverly orientated to make use of the sun. Around 4,000 years ago, people started to watch the sun throughout the year and track its position relative to the Earth as the seasons shifted. By keeping track of the sun's movements, they realised that, while the sun stayed low in the sky during the winter, shining from the south, the opposite was true in the summer. This of course varies in different hemispheres, but recognising this pattern allowed people to plan and make the most of the life-giving sunlight. Accounts of the Chinese designing homes and cities to face south stretch back to the Zhou Dynasty in around 1050 BCE.[3] South-facing buildings (in the northern hemisphere) benefit from the sun's warmth and light in the winter, but stay cool during the summer. In today's world, efficient building orientation has fallen by the wayside, as people increasingly crowd into cities. Instead of using nature, we rely on fossil fuels to heat and cool homes. South-facing properties are still sought after, but we seem to have forgotten why this is a valuable quality to begin with.

The ancient Romans went a step beyond building orientation and mastered the art of crafting transparent glass windows, which could be used as solar traps much like a greenhouse, with temperatures more than ten degrees Celsius warmer on the inside compared to the outside. This greenhouse effect was not only used for homes, but also in the famous public baths. Romans liked their baths and steam rooms extra hot, and that heat came from burning wood, often entire tree trunks (some even imported from France and North Africa), to keep up with a habit that was almost a weekly staple for some.[4] To somewhat alleviate the burden of burning trees, bathhouses were built with large south-facing windows to capture heat from the sun.

Ancient solar science and arts like burning mirrors, building orientation and trapping solar heat behind glass underpinned the development of solar energy over the following centuries. In the late 1700s, an exciting innovation called a 'solar motor' emerged, based on the idea of using a series of mirrors to concentrate the heat from the sun onto a container, causing the water inside to boil. Once the water turned to steam – think of the steam that gushes out of a kettle – the force of the steam could be used to physically push a motor, or a turbine, which in turn could generate electricity. But despite their potential, these solar motors were soon eclipsed; they were unlucky enough to coincide with the start of the Industrial Revolution and the age of coal. It was another missed opportunity for the growth of solar energy.

As Swedish-American solar pioneer John Ericsson foresaw in 1888: 'until the coal mines are exhausted [solar's] value will not be fully acknowledged'.[5] He was mostly right – but it was not only coal. Oil and gas soon followed, hampering wide-scale adoption of solar energy. It's staggering to imagine how different the world might be today if humanity had taken a different route, and clean solar energy had been the resource that powered the Industrial Revolution. Instead, the quick and easy option was taken, stunting the growth of the solar energy industry. Even today, while there are some large-scale solar projects in the world, solar is still in the process of establishing itself as a household name, despite its rich and long history.

The concept of a solar motor has survived to a small extent in some places. Spain leads the way in concentrated solar power generation, with around fifty of these curious power stations in the country.[6] Each station is a strange hall of mirrors, where reflective surfaces are carefully positioned to reflect the sun's rays onto strategically located towers, boiling the water inside them to generate steam. When night falls, these power stations can continue to generate electricity from the sun in complete darkness by ingeniously storing extra heat from the daytime in a metal tank full of molten salt fluid. It appears simple on paper – concentrating heat to boil water – but in practice, these power generators can be complicated, and expensive compared to their fossil-fuelled cousins. Unfortunately, cheap fossil fuels have not allowed concentrated solar power technology to compete, but there are some promising developments on the horizon. A wave of projects is swelling up in the Middle East, Africa and China, indicating that concentrating solar power could play a part in the energy transition.[7]

*

My annual pilgrimage to the Edinburgh Fringe, an arts and comedy festival that takes place every August in Scotland, started in 2010 after I spent a month working for a comedy club there. Street performers, musicians, actors and comedians collide for an entire month, turning the scenic city upside down. I instantly fell in love and returned year after year to enjoy the talent and chaos. I have become, if anything, a little overfamiliar with the train journey from London to Edinburgh, zooming past sheep grazing in luscious green fields set against moody skies. And, ever the engineer, I can't help doing some power-station spotting en route.

The fossil-fuel power-station skyline along the way has remained quite constant, but in recent years I have noticed another change in the energy landscape from my train window – some of the sheep now graze between row upon row of solar panels that catch the sun's energy and convert it to electricity.

Unlike orientating a building correctly to capture heat from the sun, or concentrating solar power to boil water to drive a turbine and generate electricity, the solar panels clinging to rooftops or dotting fields miss out the heat capture step. The energy is converted directly to electricity. This seemingly magic but now commonplace process all comes down to something called the photovoltaic effect, a miracle of physics that was discovered unexpectedly.

In 1839, a curious young French physicist observed a strange phenomenon. While boys his age probably pursued more conventional hobbies, at nineteen, Edmond Becquerel experimented in the laboratory his father worked in – an eminent scientist himself. At a time in history when the scientific community was busy exploring the new field of electricity, one of Becquerel's experiments involved dipping two electrodes in an acidic solution, and then shining light on one of the electrodes. To his surprise, he observed that this contraption generated tiny amounts of electricity. He didn't know why this was happening, nor did he realise that he had created the basics of the photovoltaic cell. Today, these cells are the basis of solar panels; anywhere between 32 and 144 connected photovoltaic (or PV) cells make up one solar panel, and a collection of connected solar panels make up a whole solar PV system.[8] At the time of Becquerel's experiment, little was understood about the nature and properties of light, making his discovery not only groundbreaking, but rather mysterious. And initially, scientists weren't quite sure what to do with it. But curiosity about photovoltaics started to bubble up from under the surface. Experiments by other scientists in the 1860s took Becquerel's work further – it transpired that shining a light on the element selenium could start an electric current. While Becquerel's work was with a liquid, these experiments showed that, unexpectedly, light could also cause a flow of electricity through a solid material. Rather than a messy, liquid-filled electricity-generating device, a solid device could also do the job.[9]

From the lab into the real world, in 1884, American inventor Charles Fritts constructed a very early version of a solar PV system on a New York rooftop, which did not look too different to something you might find today. Fritts had a vision that this new technology would compete with the newly introduced coal-fired power plants that had started generating electricity only a couple of years earlier. Had this technology taken off, perhaps coal would have been abandoned and we would be in a very different world today. But Fritts's solar array was inefficient, and only converted about 1% of the energy it received from the sun to electricity.[10] (To put this into perspective, the coal- and gas-fired power stations I have worked at convert about a third to half of the energy in fossil fuels to electricity.) Not discouraged by the inefficiency, and instead seeing potential, Fritts sent a solar panel to German electrical engineer Werner von Siemens, who was so impressed that he presented it to the Royal Academy of Prussia, declaring it as an example of, 'for the first time, the direct conversion of the energy of light into electrical energy'.[11] The science fiction of the day had become a reality.

It still took a long time for scientists to figure out why this technology actually worked; for decades, they only knew that it *did* work.[12] But in 1905 Albert Einstein published a series of papers showing that light has a special property – it carries packets of energy. He called these energy nuggets 'quanta', but today they are known as photons. If photons hit semiconductors (materials that only partially allow an electrical current to pass, like selenium and silicon), these packets of energy get absorbed by the electrons in the material. This gives enough energy to some electrons in the materials to enable them to break loose – give them the right conditions, and they will form an electric current.

Each of the individual solar cells that make up a solar panel has positively and negatively charged semiconductors sandwiched together to create an electric field – the right conditions for electricity to flow. Once loose, the energised electrons are forced to drift towards the positive end of the cell, generating an electrical current – this is the photovoltaic effect.[13] The loose electrons eventually hit a metal plate, completing the conversion of sunlight to electricity.

While Einstein's revelations, which built upon the work of others, finally clarified how the photovoltaic effect worked, unfortunately they didn't improve the outcomes for generating electricity from solar PV; the efficiencies remained too low to justify its use. But understanding how something works is the first step in being able to think of ways to improve it; this wasn't a dead end for the photovoltaic effect. A huge step forward came when the material of the semiconductor in the solar cells was changed from selenium to silicon, a happy side effect that spawned out of another important product: the silicon transistor.

A silicon transistor is the principal component of pretty much every electronic device we use today, from laptops to mobile phones and everything in between. First brought to life in the 1950s, the silicon transistor not only changed the face of electronics, but also led to silicon solar cells that turned out to be five times more efficient than the original selenium cells: an enormous leap forward that reopened the door for solar PV technology.[14]

But if we've had this technology since the 1950s, then why, you might wonder, has it taken so long for solar panels to take off? In 1954, Bell Laboratories (a reincarnation of a research centre co-founded by Alexander Graham Bell back in the 1800s) unveiled a small model Ferris wheel that was powered entirely by light shining on a panel of cells. While this captured the imagination of the media and public, and technical progress continued, the reality was that, at the time, solar cells cost an astronomical amount of money to make; it was calculated that an average household in 1956 would have to spend over USD$1,000,000 (over USD$10,000,000 in today's money) for a solar array big enough to power a home.[15]

Salvation for solar eventually came from the skies. A top-secret US government programme with deep pockets provided the perfect application for solar cells: an artificial satellite. In 1958, the Americans launched the satellite Vanguard I into orbit. It carried the usual chemical battery alongside the newly added six solar cells – just in case they failed. Altogether, the solar cells could generate about a watt – a tiny amount, but enough to send information about the composition of the atmosphere back to Earth. Nineteen days later, the *New York Times* announced: 'Vanguard Radio Fails to Report; Chemical Battery Believed Exhausted in Satellite – Solar Unit Functioning.'[16] Vanguard I sent information back to Earth for six years thanks to its solar cells – a massive improvement on chemical batteries, which only provided power for a few weeks since it was impossible to recharge or replace them in space. The Soviet satellite Sputnik III also successfully went into orbit with solar cells aboard, and space scientist Yevgeniy Fedorov predicted in 1958 that 'solar batteries will ultimately become the main source of power in space'.[17] He was spot-on – the International Space Station's shape is dominated by the futuristic-looking solar panel wings sticking out of it, and many forays into space are powered by solar panels once they leave the Earth's atmosphere.

But while solar cells became essential in grand space-based projects, they still had a long battle to fight on Earth. They slowly expanded into a niche in hard-to-reach areas, where recharging chemical batteries was unworkable or connecting to the main electricity grid was impossible, such as navigational lights on offshore oil platforms used to warn passing ships of the structure in the way.[18] Using solar systems rather than non-rechargeable batteries that had to be frequently swapped out saved time and money.

Solar systems also meant electricity could be introduced to remote locations. In the 1970s, it was used to bring telecommunications to people living in rural Australia. The technology needed to amplify and transmit radio and telephone signals requires a reliable power supply, which was traditionally provided by a diesel generator. But a generator needs constant maintenance and filling up with diesel – a problem in a rural place where the population is sparsely distributed across vast expanses of land. Using solar panels transformed life for Australians in remote areas, who no longer had to wait for news tapes to be flown into local stations, which had meant hearing the news hours or even days after the rest of the country. By the mid-1980s, solar PV systems became the go-to for remote telecommunications systems.[19]

While the use of solar flourished in these niche areas, it still struggled to become a mainstream way of generating electricity, unable to compete with the low cost of fossil fuels, and the complete reliance on and overconsumption of them. Following the Second World War, electricity and gas companies in the US encouraged their customers to use as much gas and electricity as possible – selling huge amounts of these products meant that companies could make money as fast as possible for a quicker return on the investments made to build power stations and power lines. Americans were encouraged and pressured with advertising campaigns to buy more electrical appliances, and the electricity and gas companies made it cheaper to use more of their products with their tariff structures, making overconsumption an easy choice.[20] This short-sighted approach completely overlooked the limited resources on the planet, and the growing impact of fossil fuels on the climate. The momentum for solar dimmed in the US as the country took a much larger interest in nuclear power, and public funding for solar research was slashed.

Outside the US, however, others took a different view. In 1990, the German government launched the 'thousand-rooftop' programme to encourage people to put solar panels on their rooftops. The government offered to cover most of the cost, and ended up subsidising the installation of over 2,000 solar systems, with a total generation capacity of about 5 megawatts[21], which today would cover 60% of the electricity needs for the 2,000 homes that had solar systems installed.[22] In the mid-1990s, Japan followed suit by offering generous subsidies to those willing to put solar panels on their roofs; nearly 400,000 had taken up the offer by 2004.[23]

The support of these governments created a market for solar PV panels, and the mass production of panels has brought their cost down more – and faster – than most people ever expected. The panels used on the Vanguard I satellite cost thousands of dollars per watt to make; today they cost less than a dollar per watt.[24, 25] Nearly 200 years on from Becquerel uncovering the starting point of this technology, converting sunlight directly to electricity is at last a viable alternative, even for grey-skyed places in northern Europe; installing new solar electricity generation today can cost less than new fossil fuel electricity generators.[26] Solar PV's rollercoaster ride appears to be reaching the end of the twists and turns, and countries bathed in sunshine can and do take advantage of this now-affordable technology.

*

Solar panels glimmer and shine like a sparkling sea in India's Thar desert, stretching as far as the eye can see. The enormous Bhadla Solar Park (an area the size of almost 8,000 football fields) reached its full generating capacity of 2,245 megawatts in 2020.[27] It is one of the largest of its kind in the world and living proof of the enormous potential for places guaranteed many hours of sunlight every day. It makes Germany's 5-megawatt 'thousand-rooftop' programme look miniscule in comparison.

This oceanic-size solar park is too large for a single company to take on the development, construction and maintenance of the whole area, so instead the Indian government divided up the land and invited multiple companies to build sections of the solar park. One of these companies, Rays Power Experts, is run by Rahul Gupta, an entrepreneurial civil engineer who spoke to me from his office near Jaipur, the capital of India's north-western state of Rajasthan.[28]

Rahul had started a solar energy company back in 2010, when solar was new in India, and the company evolved alongside the industry. He described life in India as 'more colourful' compared to many places he has travelled to, and said the country is 'growing fast; it is the land of opportunity'. When I told him I am originally from Iraq, he surprised me by saying 'Wow! Beautiful place. I really want to go there,' rather than the usual look of pity and questions about war and politics I am used to.

When they were given the opportunity to be part of Bhadla Solar Park, Rahul and his colleagues engineered and built the capacity to generate 140 megawatts of electricity – a small portion of the overall solar park, but significant, nonetheless. At the time, in 2017, the solar panels were imported from China; 'We did not have the manufacturing capacity,' Rahul told me, referring to India's solar industry. But this has changed dramatically. 'In fact,' Rahul added, 'we have no import now in India.' While this is partly because of large import duties imposed by the Indian government, it also shows how quickly the country is embracing solar.

The Bhadla project is only one of around 250 solar projects that Rahul and his colleagues have completed in India. For most projects, they play the role of project developer, taking up a large piece of land, and getting all the permits and power cables in place. Once the way has been paved, they then allow others to come in and build the actual solar park. This is particularly useful for businesses operating from outside of India, or companies that know how to build solar but don't know the ins and outs of getting the ball rolling on a project like this. When scouting areas for new solar projects, Rahul looks for land that is not suitable for anything else. The Bhadla Solar Park itself is in Rajasthan's scorching desert, where the sun shines for 300 days of the year.[29] 'It makes sense to use this land – nothing grows,' Rahul explained. 'I don't find it right to cut trees or displace a lot of animals or villages . . . sustainability shouldn't replace an existing civilisation.'

While this makes sense, it does come with certain logistical problems. Deserts mean sandstorms, leading to sand and dust covering the surface of a solar panel, which blocks the sun's rays and the photovoltaic effect, turning it into a hunk of useless machinery. The sand and dust have to be cleaned off manually, with one person sprinkling water onto the panel and another wiping it down with a mop – but it is hard labour, and water is a scarce resource in the desert.

To get around this particular issue, Rahul told me of a relatively new invention that is (quite literally) sweeping the desert: cleaning robots. The cleaning robots at Bhadla Solar Park come from Ecoppia, a company that makes, installs and maintains them. Amit Singla looks after operations for Ecoppia. He is an engineer by background, and coincidentally happens to be from the same home town as Rahul. Over a video call from Delhi, Amit explained to me that the cleaning of solar panels is a key activity when it comes to getting the most out of the panels.[30] Without the robots, teams of labourers work in the baking heat, managing one full cycle of cleaning all the panels every fifteen to twenty days before starting all over again.

'Whenever a sandstorm comes, everyone needs their plant [solar panels] cleaned as early as possible, so at that time the cost of cleaning will be much higher,' Amit told me. Ecoppia's founders designed an autonomous system that cleans the panels once a day, without any people or water. The robot itself is a microfibre brush that spins while moving down the solar panel, making use of gravity to brush any dust downwards off the panel. The brush looks similar to those used at an automatic car wash, and each row of solar panels is fitted with one robot, which moves along the row and down each panel.

'The robots have their own solar panel for power; this charges the batteries during the daytime, [and] that power is used in the evening,' Amit explained, meaning the robot does not take any of the electricity produced by the solar park. To avoid shading the panels during electricity generation hours, it does its job in the early evening. (I did want to ask if there were cleaning robots to clean the secondary set of solar panels in some sort of never-ending cycle.)

Half of Bhadla Solar Park's panels have robotic cleaning, and the operators of the other half are talking to Ecoppia about getting them installed. Cleaning the panels daily increases the electricity output by a few percentage points; while this is not a huge number, it does make a noticeable improvement to gigawatt-scale solar parks. From the outside looking in, Amit said that people think this is a simple technology, but there are thousands of robots performing a rhythmic dance, and each has to be monitored – advanced technology is critical. The robots are connected to and controlled by a cloud-based server, relying on wireless communications. Ecoppia's chief technology officer used to head up research and development teams for Israeli intelligence; his expertise on wireless communications has made the robotic cleaning very reliable. Echoing Rahul's message about India being the land of opportunity, Amit explained that all the robots are assembled at a manufacturing facility in India, and are sold to solar farms across the globe.

These solar parks are part of a larger vision for the country – India has set ambitious targets for solar electricity generation. Abundant sunshine works for India's energy security, and smaller scale solar has already helped millions of families in Indian villages to meet their energy needs, including lighting and cooking. Access to solar energy can get rid of the time-consuming job of collecting wood to use for cooking in smoky kitchens – a task that puts women and girls in particular at risk of health problems.

Between 2016 and 2022, India's installed solar capacity increased almost tenfold, improving livelihoods and contributing to emissions reductions.[31] This is incredible progress; if we move towards a future where solar is king, perhaps global power dynamics will flip – giving countries with more abundant sunshine easy access to vast amounts of energy and the political power that comes with it.

*

Traditional electricity systems around the world are structured around the idea of a few large power stations, strategically located out of people's way. The system has evolved like this over the years out of the requirements of fossil fuel-burning power stations; they take up a lot of space, they need cooling water from a nearby river or the sea, they create traffic from deliveries of materials and, in many cases, they spew out pollution into the surrounding area. So, despite being huge structures, they are hidden away. The concept of large solar PV parks mimics this idea of centralised electricity generation, but it doesn't have to. Solar panels are modular by design and don't pollute, so it is easy enough to install a mini solar power station for a single building's electricity needs, for example.

In 1987, ignoring the widely accepted gospel of a few large power stations producing electricity for an entire population, Swiss engineer Markus Real set out to prove that widely distributed mini solar power generation could also work. He put out an advert on the radio and in newspapers, looking for ordinary people who had a rooftop exposed to the sun for 'Project Megawatt' – a project that aimed to turn 333 buildings into 'power stations' using solar panels. The Swiss responded enthusiastically to the call, and he soon had 3 kilowatts of solar panels installed on each of the 333 rooftops, totalling 1,000 kilowatts (or one megawatt). An inverter, a special electronic device, connected each panel to the main electricity grid. The electricity utility company could interact with the electricity produced from these homes – if people produced more than they needed, they could sell it to the utility company, or vice versa, buying any additional electricity they needed at night or on cloudy days. Instead of centralised power generation heading in one direction, Real demonstrated that electricity generation can be a two-way street with many mini solar power stations on individual rooftops.[32]

The solar rooftop trend has continued growing throughout the world, and one country has embraced it more than the rest. Despite a past reputation for clinging to fossil fuels at a national level, Australians at a local level have widely adopted rooftop solar. Around one in four households in the country have solar panels installed on or around their homes, a larger share than any other major economy.[33] One of these households includes my friend Greg, who lives in the Jervis Bay area in New South Wales.[34] Greg is a mechanical engineer-turned-primary school teacher, who is upbeat by nature, but once I got him talking about solar panels, he became even more animated and enthusiastic than usual, going so far as to label himself a solar advocate.

When he and his partner Rachel bought their house, they immediately decided to get rooftop solar panels, which were installed a few months after moving in. While Greg does care about the environment, this decision was more financially motivated. The 6.6-kilowatt panels cost around USD$3,000, and are readily affordable because of a local New South Wales government subsidy which covers a third of the cost of the panels to encourage more uptake.[35] The subsidy is attractive, but the panels also save around USD$650 every year on energy bills. The panels pay for themselves in savings within five years, and have a lifetime of around ten years. If it's a sunny year, which in Australia it often is, Greg and Rachel spend less than half the amount on energy bills that they would without the panels. Any excess electricity produced that they do not use can be sold to the utility, but they receive a lower price than they pay out if they take electricity from the utility, 'so it makes sense to use as much of your own generation as possible', Greg told me. They try to do laundry and other electricity-intensive activities while the sun is shining. And of course, on the hottest days of the year, during the hottest part of the day, electricity from the sun can power air conditioning.

Although solar is powering individual homes in Australia, the country as a whole expends vast amounts of energy on its industrial sector – mining metals and producing chemicals is very energy-intensive, and heavily reliant on fossil fuels. Until the country as a whole decides it has had enough of fossil fuels, the actions of individual homeowners will not shift the dial.

Greg commented that when he visits the UK, where he grew up, it now feels like he has 'stepped into the future' because of the attitude to climate change and decarbonisation. 'Here in Australia, if you see a Tesla, everyone stares,' he says, whereas in the UK electric vehicles are becoming a common feature on the roads. It feels as if individual Australians get it, but as a society, he thinks Australia has missed a trick. There is plenty of empty land in the country, bathed in sunshine, but it is not being used for solar. While this is disappointing, there is hope – at a local level, there are schemes like the one in New South Wales setting ambitious renewable energy targets, and according to Greg, younger generations care much more about these issues.

Greg and Rachel's daughter will grow up in the sunshine, just as I did, and may form similar memories of warmth and light throughout her childhood. But there will be a difference, because she will see the solar panels on the roof of her home and associate the sun with clean electricity. From the ancient history of heating homes and bathhouses with the sun, to the development of photovoltaic cells with the help of space missions, the direct harnessing of the sun's energy is now on a promising rise. The efficiency of solar cells has improved by leaps and bounds, from the initial measly 1% to newer designs today exceeding 30% – a phenomenal improvement.[36] There is still a mountain to climb, but the road ahead looks less steep. And for places without plentiful sunshine, we can turn to other forces of nature in the mission to transition away from fossil fuels.

Chapter 4

Wind and Water

'I'M HERE TO look at a wind turbine,' I beamed enthusiastically at the Danish border control agent. She looked a bit perplexed.

'So . . . you're here for work?'

'No. I've come to see a wind turbine called Tvindkraft; it was built by a group of teachers in the 1970s,' I blathered on, wishing I had just said a simple 'I'm here as a tourist.' She seemed bemused but let me through. I told her that my mother was the next person in the queue to save her from going through the same rigmarole.

Looking out of the plane window as we descended into Billund airport, I was thrilled to see Denmark's wind energy heritage obvious even from the skies. Little three-bladed wind turbines speckled the land below, their crisp whiteness complementing the lush green land (making my 'I'm here to look at a wind turbine' even more ridiculous – it would be almost impossible to go to Denmark and *not* see a wind turbine).

The world's history of wind energy goes back hundreds of years, to blacksmiths and farmers using the energy in the wind to do mechanical work, like running farming equipment or crushing grain. By the 1930s, Denmark was ahead of many other countries when it came to wind farm development, and wind turbines had become a typical sight in the countryside. But the rise in popularity and availability of cheap coal, and subsequently oil and gas, soon crushed the growth of the industry – the same fate that solar energy encountered.[1]

As the world emerged from the two world wars, and more oil resources were discovered and extracted during the 1950s and 1960s, oil prices plummeted, giving it the edge over other fuels, including coal. At the time, oil seemed like a fuel without many downsides (a warning we should heed when we consider fuel alternatives now), though there were early scientific murmurs linking fossil fuels with global warming. In western Europe in 1955, coal provided three-quarters of the total energy used, and oil just under one-quarter. By the early 1970s, the picture had completely flipped, with oil overtaking coal, not just in Europe, but globally. By 1971, oil and natural gas provided two-thirds of the world's energy needs, and since then fossil fuels have well and truly dominated the energy sphere.[2] During the 1970s, however, the demand for oil started to catch up with the abundant supply, and many countries soon developed a dependence on oil from the Middle East and North Africa.

Things came crashing down in October 1973, when an Arab–Israeli war erupted in the Middle East. Arab oil-rich countries like Iraq and Saudi Arabia raised oil prices, and stopped selling oil to countries that supported Israel. It was the perfect recipe for panic – conflict, a lack of supply, and not knowing when oil would flow again. In many countries across the world, long queues of people and their cars formed at petrol stations, and the panic-buying caused prices to go up even more. The shocks were felt worldwide, taking countries back to the post-war years of shortages and exposing how dependent they had become on energy imports – a dynamic that still exists today.

In Denmark, the events of the 1970s fuelled the debate about alternative supplies of energy, and in particular the option of nuclear power. But a viable alternative came from an unlikely place: a group of Danish teachers who were opposed to nuclear and wanted to take the country's energy solutions in a different direction. They wanted to do what Denmark had been good at historically, by proving that renewable energy from wind could work on a large scale. Teachers at a collection of independent schools known as Tvind, including a teacher-training college, formed a team to design and build what was at the time the world's largest electricity-generating wind turbine: the one-megawatt Tvindkraft, the creation that prompted my visit to Denmark (and an awkward interaction with the border agent).[3]

In May 1975, 400 volunteers broke the ground to prepare the site for the wind turbine. A team of about thirty-five teachers, students and volunteers formed the core group of designers, builders and decision-makers. The average age of the group was twenty-one, and the youngest member was just sixteen. All decisions had to have unanimous agreement within the group, who had no blueprint or manual describing how to build one of these at this scale. Everything had to be designed from scratch.

The team felt the pressure, conscious that wind energy's very reputation hung in the balance. If they failed, wind power would not be taken seriously as a large-scale electricity-generation method, and nuclear would win out. But if they succeeded, they could start a new era of renewable energy generation for Denmark. Every detail was carefully considered, from how to mix the concrete to how to tighten the bolts. Where they lacked knowledge, they pulled in engineers and consultants to help out. About three years after the ground was broken, the wind turbine started producing electricity. Entirely community-owned and financed, the project was akin to a university for how to build wind turbines, with many members of the team going on to work in the budding renewable energy industry.

The turbine still operates today, more than forty years later. The location of the turbine is also home to seven schools and social care homes for around 120 people, complete with a football field, sports hall, music and art facilities and gardens. On balance, Tvindkraft produces enough electricity to power these facilities, selling power to the main grid when there is plenty of wind and buying when there is not. Denmark is littered with wind turbines, on land and at sea, but this one is very easy to spot – painted with red-and-white pop-art geometric patterns designed by Danish architect Jan Utzon. It is a unique structure, standing out against the surrounding landscape, like a teenager daring to be different. As my mother and I drove through the Danish countryside, the bold red and white of the turbine popped into view on the horizon. It was a bright, sunny and relatively still day; the turbine blades were lazily spinning. Allan Jensen and Britta Jensen, previously teachers but now the turbine's caretakers, welcomed me and my mother with an extensive selection of biscuits and coffee.[4]

For Britta, the key to community acceptance and success of wind power is local ownership. 'Make renewable energy a local project, involve the local people in it, give them the chance to participate,' she said. The concept of local wind co-operatives, where a group jointly invest in wind turbines, became popular in Denmark in the 1980s.[5] Denmark also has a strong decentralised energy generation trend; rather than the giant centralised power stations pumping out electricity and heat that are so common across the globe, each municipality has smaller, local power plants. This may be a way of gaining support for renewable energy projects, by offering locals a tangible stake, and it is an approach I have witnessed in other parts of the world.

Allan offered to take us to the top of the fifty-four-metre-tall turbine, though my mother opted to stay at ground level. First, we stopped off in the little control room to switch off the turbine, so it would be safe to go up. Allan pressed a combination of keys on the computer in the control room to turn off the machines and to change the angle of the blades, to stop them spinning. That was enough to allow us to get up there, where a physical blade brake can be engaged to lock them in place. From the control room, we walked across a neat lawn and past a sculpture of a giant dining table and two dining chairs, towering above me, one of many works of art added to the site for visitors to enjoy, and part of the effort to engage tourists and locals in the world of renewable energy.

When we reached the turbine, a rickety but reliable lift – a former building hoist – was our ticket up. Allan gave me a hard hat to put on, and a few moments later the lift jolted to a stop at the top of the tower. I climbed the narrow steps leading up to the nacelle – the pod that sits behind the turbine blades containing the equipment that generates electricity. I hit my head on something above me with a loud bang, and thanked Allan for the hard hat.

The space inside the nacelle was cramped, but big enough for a couple of people to work on the equipment. It was filled with a familiar power station smell of oils and grease that kept the moving parts mobile. The mechanism is surprisingly simple. Starting from wind speeds of about four metres per second, the wind spins the three turbine blades mounted at the top of the turbine tower, spinning an input shaft. A series of gears inside a gearbox increases the number of times the shaft spins per minute, so the output shaft moves much faster. The next box along is the generator, converting the kinetic spinning motion into electricity, which is carried away in cables to where it is needed.

I knew already that Tvindkraft's parts were mostly recycled but was delighted when Allan pointed out the various quirks of the turbine: the shaft came from an oil tanker in Rotterdam, the gearbox was a spare from a copper mine in Sweden, and the generator came from a paper factory in Sweden. Aside from producing renewable power, this beautiful wind turbine has given these old parts – junk, some might say – a completely new lease of life.

Tvindkraft's parts are not static. That may sound obvious because it's a wind turbine, and turbine blades are meant to spin, but the blades move in unexpected ways too, and other parts of the turbine also move to get the best out of the wind. The turbine blades' pitch – the angle of the blades – changes to capture maximum energy from the passing wind, rather than remaining at a fixed angle constantly. This ability to change angles also acts as a safety feature; if pitched to face the wind, the blades stop turning altogether, which is also how a turbine is switched off, and how Allan turned off the machine in the first place to allow us to go up there. At high wind speeds, above about fifty-seven kilometres per hour, the turbine turns itself off to protect its internal organs. The whole of the nacelle, with the blades attached to it, also moves. It rotates or 'yaws' on its axis, depending on the wind direction. All of this happens automatically with a controls system that monitors the wind conditions and adjusts the positions accordingly.

Allan pointed out that this turbine is downwind, so the wind blows from behind. Having looked closely at a lot of wind turbines, to me, the blades looked as if they were mounted the wrong way around, with the angle pointed away from the tower. This means that the wind hits the back end of the nacelle first rather than the face of the turbines. Unlike some of my accidental mistakes with flatpack furniture, this was a very intentional decision, as it was the safest option at the time. The blades were made from very strong fibreglass, an innovative design, so the team did not know how much they would flex in the wind. If the turbine had been built the 'correct' way around, the blades would have been angled towards the tower, and they could have flexed enough to hit the structure. A wind-measuring device mounted on the back end of the nacelle measures the wind speed before it gets to the blades, giving a more accurate reading and more time for the turbine to react if it needs to shut down. And finally, when built this way, if the rotation of the nacelle fails, it is in a safer position.

Climbing a small metal ladder through a hatch at the top of the nacelle, I popped my head out into the fresh air beside the turbine blades. Looking out at the landscape, I could see many more wind turbines, watched over by this multicoloured matriarch over the years, a symbol of hope for a renewable energy future. The reason I had wanted to see Tvindkraft for myself is that it is thought to be the oldest operating wind turbine; it is a true pioneer and has also played a political role, showing what can be achieved through local co-operatives. It cemented Denmark's position as an international leader for wind energy, a position the country continues to hold.

Back in our apartment that evening, my mother video-called her friends in Baghdad. 'Do you remember the war in 1973? When oil stopped moving to the West?' she asked, and then went on to rave to her friend about Denmark and wind energy. A medical doctor with an obsessive interest in war and politics, up until now she had shown no curiosity about wind energy. The value of seeing the turbine and hearing its story first-hand from the caretakers had completely converted her, and convinced me to encourage others to see it for themselves. Opening up energy infrastructure to the public – and sharing the stories and quirks behind these seemingly lifeless structures – has to be one of the best ways to get people on board.

*

Some people think wind turbines are an eyesore, but I love seeing the giant blades gently whooshing in endless rotation

– whether on land or at sea. To me, they are beautiful and functional structures. They also give me hope that, in combination with other ways of generating and using energy, we could start to undo the climate damage unleashed over the past years and have a fresh, sustainable future.

The basics of today's turbines are not too different to Tvindkraft: a tower, three blades, and a nacelle containing the shaft, gearbox and generator (although a few modern models are direct-drive, without a gearbox). The components are more advanced these days, and the turbines are bigger, of course. And increasingly, turbines are being built out at sea, where they are less visible to local communities, and competition for space isn't as fierce. Tvindkraft was not my first encounter with a wind turbine; a couple of years earlier, I had managed to get very close to an offshore wind farm called Rampion. Anyone can book a boat tour of Rampion offshore wind farm, thirteen kilometres off the coast of Brighton in the UK. You can even see the 116 white wind turbines from Brighton's coast – tiny white sticks on the horizon.

On the day of my visit, the combination of a sunny day, being on a boat, and approaching the offshore wind turbines fired up my rather unconventional excitement for energy infrastructure. It was my first time getting so close to an offshore wind turbine, and the physical size came as a shock. I had looked up the numbers (140 metres tall to the tip of the blade), but found it difficult to picture that in real life. As I looked up, two workers were suspended from ropes, working on one of the blades. Their figures put the size into perspective – two tiny dots against the bright white surface of the blade.

Each individual turbine can output a maximum of 3.45 megawatts of electricity – for comparison, Tvindkraft outputs just under a megawatt. In the right wind conditions, when all the turbines are spinning at full pelt, the farm can generate 400 megawatts, enough electricity for around 400,000 homes. Out at sea, turbines can be physically bigger, since there is less concern about the structures blocking views. This, coupled with typically stronger winds, means fewer turbines are needed at sea than on land to generate the same amount of electricity. At sea, wind speeds and directions do not vary as much as on land either, giving more reliable power generation overall.

When the wind pushes the turbine blades, that energy is converted to electricity, which is moved to land in a metal cable, wrapped in plastic insulation, and buried just below the surface of the seabed. According to our captain, who was a fisherman, the wind farm has had a positive impact on biodiversity. Trawling for fish is not allowed in the area, so populations of sea creatures down there have thrived, protected by the giant sentinels.

Despite all the advantages, wind still only plays a tiny part in the energy sector – in 2021, only 6.7% of the electricity generated worldwide came from wind, although it's a step up from just 0.2% in 2000.[6] The potential is enormous and lots of countries already have big plans to build wind farms to generate clean electricity; Europe and the US have large targets and, as part of decarbonisation plans, China is expected to generate 700 gigawatts of wind power by 2030 (the equivalent of 1,750 Rampion wind farms).[7] If all of these projects happen, it would go a long way to decarbonising electricity supply.

WIND AND WATER

Offshore wind turbines, like the ones at Rampion, are fixed into the seabed. The tower that the blades are attached to is stabbed into the ground, keeping the whole thing supported and in place. As a result, these fixed foundation turbines can only be installed in quite shallow seawater. Any deeper than about fifty metres, and it becomes technically difficult and expensive to fix the tower to the seabed. But the further out to sea you go and the deeper the water gets, the more consistently the wind blows. It may seem a shame to miss out on this potential source of renewable energy just because we can't install the foundations of the turbines into the seabed. But, of course, there is always another way, and the ingenuity of engineers has delivered once again.

About ten kilometres off Karmøy in Norway, in 200-metre-deep water, a 3.6-megawatt turbine stands so steadily that it's difficult to believe the structure is actually floating on top of the water. When I went to Norway for a conference about floating offshore wind, I got the chance to visit it in person. I was so keen to catch the first glimpse that I spent the entire forty-minute boat journey sitting in a white plastic chair on the top deck, ignoring the biting cold (despite it being June), my eyes fixed on the horizon. When the turbine finally came into view, I could make out the white tower with the three blades topping a yellow tetrahedral structure, made from tubes of steel, similar in size to the Rampion turbines. At this point, I shifted to the front of the boat and leaned against the metal railing for a better look, like a child in an aquarium pressing their face against the glass. I was blown away by how still the large structure was, easily resisting the movement of the current. Although the sea was calm, the boat I was on bobbed up and down in the swell like a gentle fairground ride. The turbine, though, was weighed down by a keel – a massive subsea structure. Like the large mass of a giant oak tree's roots that keep it grounded, the floating turbine's keel, anchors and moorings work together to keep it sturdy and balanced.

The excitement of being there kept me glued to my spot, taking in the view with awe for the next thirty minutes or so – however, I paid for this as we started to sail away from the turbine and the seasickness hit. I returned to my white plastic chair on the deck and buried my head in my hands until we got back to shore. Sickness aside, I was still elated to have seen the world's first full-scale demonstration of a floating offshore wind turbine, a partnership between Dutch, German, Japanese, Danish and Norwegian energy companies.

Floating wind could be the next frontier, unlocking vast areas of the sea for harnessing renewable energy. This development makes the humble Tvindkraft look tiny in comparison, but without those pioneering beginnings, none of it would be possible. I remembered my conversation with Britta at Tvindkraft, who had told me she was disappointed with Denmark's energy mix. 'Denmark could easily today have had 100% renewable energy,' she had said, describing the country's 50% renewable energy as 'embarrassingly low'. My mother had been surprised by this level of ambition, repeating this to her friends in Baghdad on her video call. She also shared a fact I had told her earlier that day – Norway's electricity comes from 100% renewable energy. 'What about their oil and gas?' one of her friends had asked. 'Oh, they export that, and use the profits to make life better for Norwegians!', my mother eulogised. 'They're just like us, then,' her friend replied sarcastically.

*

Norway's secret to achieving 100% renewable electricity isn't down to wind, though. Instead, they have successfully exploited another abundant natural resource: water. The motion of water as the tide changes or as waves traverse the sea, or even a large lake, can be harnessed for energy, but by far the most prolific type of electricity generation from water is through using a dam or similar structure to control and harness its flow.

More than 90% of Norway's electricity comes from hydropower. Just as ancient civilisations used windmills to power machinery and crush grain, water mills were utilised in a similar way. Instead of laboriously grinding grain by hand, or painstakingly moving machines with people or animal power, the flow of water pushing against a water wheel generated motion. This is very similar to the way a wind turbine works, using the motion of the water rather than wind. Until the late 1800s, that was as far as it went; the turning force of the water wheel was harnessed and directly transferred to the turning force of the mill, or grindstone. But taken one step further, that same motion can be used to spin a turbine to generate electricity.

The first major use of hydroelectric power was in 1882, when the president of a paper mill pushed for an installation of a power station on the Fox River in Appleton, Wisconsin, US. It produced enough electricity to light the paper factory and a couple of nearby buildings. In today's world, it would have been enough energy to light more than a thousand LED bulbs.[8] That doesn't seem like much now, but it kicked off the appetite for hydroelectricity, just as Tvindkraft would later spark an interest in and demand for wind-generated electricity. By 1920, a quarter of all electricity generation in the US came from the flow of water.[9]

Today, hydropower is responsible for the biggest share of renewable electricity generation worldwide – partly thanks to its long history, and partly thanks to the fact that places with a good flow of water can take advantage of this natural gift. Wind and solar are starting to catch up, but hydropower stations have steadily supplied about 7% of the world's total energy demand, not just electricity, over the past few decades – with China, Brazil, Canada, the US and Russia being the largest producers of hydropower in the world.[10, 11]

Norway's hydropower prowess is impressive in the context of providing the country with the vast majority of its electricity, but it doesn't feature on the list of largest producers because of its relatively small population – around five million inhabitants in all. Its hydropower capacity is dwarfed by China, which has more than ten times the installed capacity, providing about a fifth of the country's electricity demand for the 1.4 billion people living there.[12, 13] China is also home to the world's largest electricity generator, both renewable and non-renewable – a hydroelectric facility on the Yangtze River.

Slicing through China, the Yangtze River has been the lifeblood for many civilisations rising and falling over the course of thousands of years. Melting glacial waters from the Tibetan plateau snake 6,300 kilometres through mainland China, past major industrial cities like Chongqing, Wuhan, Nanjing and Shanghai, before spilling into the East China Sea. About halfway up the river, three gorges line up side by side, stretching over 300 kilometres. Overlapping cliffs and lush woods have seen generations of inhabitants come and go, hearing whispers and rumours about hydroelectricity for nearly a hundred years, before finally witnessing the construction of a giant: the Three Gorges hydroelectric dam.

The idea of a dam across the river was originally imagined by the first president of the Republic of China, Sun Yat-Sen, in 1919. He dreamed of a series of dams that would harness the natural energy of the rivers to provide power for the recently established vast republic.[14] In 1924, suffering with cancer just a year before his death, Sun Yat-Sen painted a visionary picture in a public lecture: 'If the water power in the Yangtze and Yellow Rivers could be utilized by the newest methods to generate electrical power ... When that time comes, we shall have enough power to supply railways, motor cars, fertilizer factories, and all kinds of manufacturing establishments.'[15] As well as powering industry, he believed a dam could control the devastating floods the country had experienced about once a decade for the past few thousand years at the hands of the river. These floodwaters could rise up to eighteen metres in a single day; enough to reach the sixth floor of a building. Not only that, but the waters were rough, with rapids, whirlpools and rock falls smashing and swallowing up any small passing boats. Building a dam would improve the shipping route, calm the waters and create a safer passage between the industrial cities along the way.[16] Sun Yat-Sen's vision seemed to benefit everyone – yet for decades, wars and politics got in the way of making it a reality.

In 1954, Mao Zedong, founder and leader of the People's Republic of China, attempted to resurrect the hydropower project. He was searching for a solution after a long rainy season had led to more catastrophic floods around the middle stretch of the Yangtze, killing around 30,000 people. But the tribulations of the Great Leap Forward and arguments about where to build the dam, along with other reasons, hampered the project. It wasn't until 1994, seventy years after the dam was first conceived, that construction finally started, and in 2012 the world's largest power station became fully operational, representing China's emergence as a superpower on the international stage.

The amount of electricity that can be generated from a hydroelectric power station depends on how much potential energy is available, which in turn depends on gravity, the mass of water, and the height it is dropped from. The force of gravity is constant on our planet, but more electricity can be generated with a larger amount of water and a longer drop. At the top of the Three Gorges Dam, a reservoir holds back a huge amount of water, before it is allowed to drop 185 metres – about twice the height of the Statue of Liberty in New York – through giant concrete tubes resembling massive water slides towards the generators placed at the end of the tubes. The force of the dropping water spins thirty-two separate turbines, with a capacity to generate a whopping 22,500 megawatts. That's eleven Ratcliffe-on-Soars, the large coal-fired power station I once worked at.

At such a great height, and stretching 2,000 metres across the width of the gushing river, building the structure was truly a feat of engineering – before any work could even begin, the water in that section of the river had to be physically moved out of the way. An artificial canal was dug to bend the Yangtze away from its natural route, like a diversion for roadworks. When that stage was complete and the river finally moved into the diversion canal, fireworks went off – thousands of spectators cheered to celebrate, and nearby ships joined in with their horns.[17] Two temporary dams were built on the riverbed to stop water flowing through to the section where the actual Three Gorges Dam would be built, and the riverbed in between was dried out, ready for the concrete to be poured and for construction to begin.

The designers had to think about other users of the Yangtze, a river that also serves as a busy shipping route. A huge concrete wall in the way of the ships and boats would have been quite unpopular, so a ship lock that allowed ships to move up and down the river, across the dam, was installed. Ships enter the lock, the gates close, and water floods in, lifting the vessels up to the next level. Five tiers of locks eventually get them to the other side to continue their journey, but it takes time. For smaller boats looking for a quicker option, a thirty-minute elevator to get up or down the dam is also available.[18]

While my interest and attention naturally drift towards the technicalities of the power station itself, the sheer act of constructing this one came with so many other challenges and consequences. For a start, adding a dam to a river causes flooding. Imagine putting up a wall in the middle of a bath tub while the tap is still running – that section of the bath will fill up and overflow in all directions, not just into the next section of the bathtub. In practice, this meant that an estimated one to two million people in hundreds of villages, towns and cities upstream of the planned dam were evacuated and relocated. Areas were permanently flooded, swallowing up archaeological sites, disturbing natural habitats, and potentially increasing the risk of landslides.[19]

Dachang, a town up in the mountains, was one of the few remaining communities dating back to the Ming Dynasty (1368 to 1644), and had been a quiet haven for families and their respected ancestors. Ancient city gates, stone roads and houses all survived the test of time, and in 2003 the Chinese Relics Protection Department decided to save at least some of the town before it got engulfed in the flooding of the Three Gorges Reservoir. Thirty-eight houses were chosen, carefully marked and taken apart, brick by brick and tile by tile, before being moved and reconstructed at a safer site, five kilometres away. But many historical sites were not so lucky, and neither were the 10,000 inhabitants of Dachang, all of whom were forced to relocate.[20]

The Chinese government built new towns and offered new housing for the relocated people, but being forced to leave one's ancestral home is a bitter pill to swallow. Some adapted well, and perhaps even benefited from the relocation, but others did not. Rural inhabitants, who had made their living for generations through farming the land that would go on to be flooded, found it difficult to adjust. Some were resettled and given plots of land to farm, but these were of poorer quality and different to the environment they were used to. Some people were relocated far away from the Three Gorges area, and found themselves isolated and struggling to communicate, their dialect incomprehensible to their new neighbours.[21]

Though the dam is a huge engineering accomplishment, unlike anything else on the planet, there are consequences that go hand in hand with such a massive undertaking. The scale of everything involved was enormous, from the concrete poured to the lives affected. The evacuations and relocations were complicated projects in and of themselves, and creatures in the river were affected by the huge barrier placed in their habitat. Fish that could previously move freely now have to navigate a giant structure; dam builders do think about these things, and design fish ladders to guide fish around the dams, but this doesn't compare to their natural habitat.

The reservoir itself means emission-free energy, but it comes with other concerns, including the release of methane, a potent greenhouse gas that contributes to global warming. Low oxygen levels at the bottom of the reservoir can create the right environment for bacteria to decompose organic material, like the trees and grasses that lived there before the area was flooded. The methane produced from the decomposition bubbles to the surface, lingering in the atmosphere. However, despite all the downsides, electricity generated from hydro schemes has far lower greenhouse gas emissions when compared to fossil fuel-powered generation – tens of grams of carbon dioxide per kilowatt hour of electricity, rather than many hundreds.[22, 23] The low-carbon electricity generated undoubtedly helps to decarbonise China's energy system, but some engineers from China and elsewhere believe that multiple smaller dams along the Yangtze tributaries would have achieved the same effect – generating as much power, and providing effective flood control just as well as the Three Gorges Dam, with less environmental impact and displacement of people.[24]

Looking at the decade ahead, China is expected to remain the single biggest hydropower market. But globally, it is thought that nearly half of the economically possible hydropower potential out there is untapped, and others are catching up with China's progress. Hydropower projects are expected to spring up across India, Laos, Nepal, Colombia, Argentina, Turkey, Canada and others, converting cascading water into clean electricity.[25]

Looking back on it now, it is difficult to determine if the giant Three Gorges Dam was in fact the best solution, or if it was simply a display of power. Does the clean energy generated make up for the relocation of communities, loss of natural habitat, and change to the landscape? Or could there have been a better solution? New hydropower projects will face similar issues, but perhaps not at the same scale. Difficult decisions and sacrifices will undoubtedly have to be made in exchange for low-carbon power.

*

As an engineer, I have learned a huge amount over the years; having personally contributed to the fossil fuel industry and felt the accompanying excruciating guilt, I now work in renewable energy. I am not the only one changing; there is a general shift in attitude as more and more people wake up to the reality of the impact our energy systems have on the planet.

There is a slow, but nonetheless encouraging, push towards more renewable energy sources. It is a visible change – I now see more solar panels on rooftops, large solar farms on the ground, and wind turbines on land and at sea. As renewable energy blossoms, it is tempting to sit back and breathe a sigh of relief, waiting for the transition to roll in. But just because energy is renewable doesn't automatically mean it is obtained responsibly, so it's important to keep a close eye on and learn from the good examples set by the likes of Norway and Denmark, especially where communities have taken the lead and thought about the longer term.

Renewable energy is also not squeaky clean – the word 'renewable' does not necessarily mean that zero greenhouse gases are emitted in the process. Carbon dioxide emissions can come from the manufacture of the materials and equipment needed to build the infrastructure, and during maintenance. Fossil fuel-powered boats often bring spare parts and people to offshore wind farms. Renewable energy generation is a huge step in the right direction, but there is a long road to true sustainability.

As with nuclear energy, responsibility for renewable energy infrastructure extends beyond the here and now – wind turbines are designed to last for about thirty years, after which they are taken apart and disposed of. Tvindkraft set a good example by reusing components from other industries, but this is not the standard practice. Amongst the tangle of components that make up a wind turbine, the turbine blades may seem quite harmless, but they are unwieldy and not yet widely recycled. Most are made from a mixture of materials glued together to give them the correct strength and flexibility, but that blend makes turbine blades hard to recycle as there is no practical way to separate the materials from each other. In most places, for now, these giants are cut down into smaller pieces and buried in landfills, or even burned.[26]

But there is hope on the horizon. Germany, Austria, the Netherlands and Finland have banned the landfilling of turbine blades, and some turbine manufacturers are introducing a design that allows separation of the blade's different materials at the end of its life, for reuse in other applications.[27, 28] Thinking about these issues before even designing the turbine blade is the attitude needed to create a more sustainable planet. Consider the full lifespan of a power generator, from how it is constructed and how it generates energy throughout its life, to what happens to the materials at the end.

Responsibly installing renewable energy generation and figuring out what to do with the equipment at the end of its life is only the beginning of this story. To benefit from and use any type of energy, renewable or fossil-fuel generated, it first has to be moved from the source to its consumer. A farmer can grow potatoes, but they are no use if left to rot on the field – they have to be picked, moved and made available for people to eat.

In the future, whether energy is captured from the sun, wind, water, or by burning fossil fuels, or by nuclear fission or fusion, we will still face the same mammoth task – moving that energy around the world. Across the world, pipes lie hidden underground filled with natural gas flowing through them, giant ships carry loads of oil or coal across international bodies of water, and overhead cables carry electricity across borders.

It is a bespoke yet global system of moving energy – bespoke for the local context and communities, and global because it is a highly interconnected system. Problems on one side, like the oil crisis of 1973, seep across to all corners of the system and those connected to it. Much like anything in life, nothing exists in isolation. To truly understand energy and its future, we need to see the system in all its complex, global, interconnected glory. After all, it is this system that holds up our lives as we know them.

PART 2

MOVE IT

Chapter 5

Sea and Land

I WOKE UP in the middle of night, shivering and freezing cold. I deeply regretted not spending more time fiddling with the heating system of my thin-walled Portakabin. It hadn't seemed important during the day, when the heat from the glaring sun made the air temperature a balmy seventeen degrees Celsius on this winter's day. But the temperature dropped by more than ten degrees as soon as the sun set, dramatically marking its absence. Of course, it's cold at night in the desert. How could I forget this rather basic fact? My tired and frozen brain, having struggled and failed to figure out the system for turning on the heating, gave up on the problem, and so I shivered the rest of the night away.

The sun returned in the morning, melting away my exhaustion as the bright light bathed the sea of sand around me. Over the next ten days, I got used to the temperature flip, just as I adjusted to the strange way of life inside this gated compound, complete with armed guards. This work trip to Hassi Messaoud in Algeria was early on in my career, during my time in oil and gas. I spent some time with the exploration team, who were drilling holes down into the depths of the desert in search of fuel. Most of us stayed within the walls of our compound, doing desk-based work and flying out in small airplanes into the desert where the actual drilling was happening – an even more remote location – only when needed.

In Algeria, I felt a constant mixture of excitement and guilt. It was exciting to visit a new place and meet new people, but this was dampened by the guilt that came with my parents' horror at my being there. They had gone through years of sacrifice to get our family out of Iraq and, from their point of view, I was willingly placing myself back in a similarly dangerous country. More horrifying to them still was that, at the time, I wasn't viewing this as a one-off trip – I was considering working there on a more permanent basis. Needless to say, I did not mention the ageing airplane I boarded for a day trip to the oil rigs in the desert, the barbed wire topping the walls of the compound I stayed in, or the number of armed guards patrolling the site.

Despite my parents' worries, my ten-day trip was perfectly safe; I felt welcomed by the hospitable Algerian culture, and got a kick out of attempting to communicate in Arabic, my Iraqi dialect almost incomprehensible in most of North Africa, where French has woven its way into the language. My parents' fears were not completely unfounded, though; a few weeks after my return to the UK, terrorists attacked an Algerian gas terminal close to the border with Libya.

The company I worked for at the time was one of many international organisations searching for oil and gas in the region, because Algeria has such large quantities of these resources below its golden sand dunes. Despite my trip being a one-off in the end (to my parents' relief), it did open my eyes to this country's energy industry and underline how interconnected our global energy system is.

From extracting fossil fuels to harnessing the energy from the wind, water and sun, the energy used every day to light our homes and power our transport systems is gathered from the world around us. But getting fossil fuels out of the ground or generating electricity from the sun is only the beginning of a very long process. Before any of this energy can be used, it must be moved from source to end consumer: from sunny, arid desert to damp and cold islands; from gargantuan dams to thriving urban centres.

Energy is not equally distributed around the world. Algeria produces more fossil fuels than it needs, so the excess is exported to other parts of the world less fortunate in the fossil fuel lottery. But countries like Japan, Malta and Cyprus, and other island nations with very little in terms of local energy resources, depend almost entirely on energy imports. Over 90% of the energy used by the populations on those islands often comes from outside of their borders, from areas that are more fortunate, at least in terms of their natural energy resources.[1] And for those countries that, by contrast, have an excess, like Algeria, the ability to move energy from one place to another has contributed to their development as nations.

Moving something as varied and seemingly intangible as 'energy' has led us to think up a number of creative, often complex solutions to the world's unequal energy distribution. There are multiple options for moving different types of energy. Solid, liquid or gaseous fuels like coal, oil or natural gas, can all be moved by sea, rail, road or pipeline, while electricity is moved through cables and wires. Selecting one of these methods depends on what is being moved, where it's being moved from, where it's going, and of course the existing availability of transport infrastructure.

Huge ships can move mountains of coal or deep pools of crude oil from one side of the world to the other. Barges floating along canals and rivers, as well as trains, historically shifted coal from the head of a coal mine to consumers. Today, small amounts of liquid fuels are moved in cylindrical metal tanks by rail or road, and a lot of natural gas is moved by pipelines. Both pipelines and cables can travel under the sea, connecting continents and bringing abundant oil to coastal nations.

I tend to associate pipelines, ships, trains and trucks mainly with moving fossil fuels, and this association is justified, since it reflects the state of the energy web today. Renewable energy from the sun, wind and water can also be traded between countries, but this must first be converted to electricity before moving through metal cables. As we travel towards a future more reliant on renewables than fossil fuels, some places will use their natural environment to generate their own energy, reducing the need for imports. But we will likely continue to see energy sharing to manage the intermittent nature of renewable energy, although the methods and routes may look different.

Moving energy resources is a truly global business, and an essential one at that. Being a part of this gigantic energy web enables countries to secure a steady energy supply and give their citizens a better standard of living. It is difficult to imagine Tokyo's brightly lit night-time skyline and technology-driven society existing without the infrastructure to bring in energy. Japan relies on imports from all over the world – from the coal that is shipped from Australia, to the natural gas that also comes to them from Australia and Malaysia – to fuel its citizens' needs and lifestyles.

Given that all of us, to some degree, rely on or are relied upon by other countries to power daily life, it's not difficult to see how critical international co-operation is. It is a delicate dance, with the impact of wars between nations radiating far beyond the point of conflict and being acutely felt by those responsible for moving energy resources across land and sea.

*

Moving liquid oil by ship across waterways has been happening since the 1850s – it is easy enough to picture pouring a liquid into a tanker, and the ship drifting from one place to another, offloading this product like some sort of international milk round. But shifting natural gas, a fossil fuel that has a shorter carbon chain than liquid oil, is a very different beast. Its very nature as a gas makes it tricky to move – imagine trying to capture and move a kilogram of the air around you to another country. This problem never bothered the early oil explorers of the 1800s, who found natural gas alongside the crude oil they dug up but treated it as a nuisance by-product. They just wanted the fossil fuels that were visible to the naked eye. Not realising the value of the gas, they simply got rid of it.

It took a long time for natural gas to establish a name for itself, and for its movement across the globe to become critical. Nowadays, natural gas is mostly used to generate electricity, as well as heat for homes and industry. Well before this, the streets of British cities in the 1800s were lit by 'town gas', a noxious gas made from coal. Town gas eventually made its way into homes for lighting, cooking and heating, an improvement on having a coal fireplace spewing smoke and carcinogens directly into people's lungs. But the gasworks that transformed coal into town gas spilled smoke and carcinogens into the surrounding air, poisoning workers and local residents. The pollution was still there, it had just been shifted out of homes.[2] Huge improvements in pollution levels came about when town gas was finally replaced with natural gas in the 1960s; burning natural gas was far better for the air quality in cities.

Using cleaner natural gas improved people's standard of health and marked a change in the importance of the substance for energy security. It was becoming increasingly important to be able to move natural gas from producers to users. The oil shocks of the 1970s that led the Danish teachers to build the Tvindkraft wind turbine cemented this position – it drove the demand for natural gas to more than double, as countries were forced to think about how to diversify their energy supplies to be less reliant solely on coal and oil. This huge increase in demand called for bigger and better ways to move gas from countries like Algeria, where it was found, to where it was used.

The key to moving natural gas is to first turn it into a liquid. In this state, it takes up much less space than in its gaseous form, like a genie squeezing into a bottle. Natural gas can be turned into a liquid with extreme cooling to below its boiling point of minus 163 degrees Celsius, making liquefied natural gas (LNG) in a process known as 'liquefaction'. In liquid form, it takes up about 600 times less space. Once the LNG reaches its destination, where it needs to be used, it can be warmed back up to its gas state in a step known as 'regasification'. In a liquid state, it can be more easily transported over large distances, most commonly by sea on a ship. A ship carrying gas as a liquid sounded nonsensical when I first came across this idea, but it's not so strange once you stop to think about it. Smoke machines, often used in concerts and theatre productions, use liquid nitrogen or carbon dioxide squeezed into cannisters that appear to be far smaller than the amount of gas you see spilling out over the floor of a stage – they use exactly the same principle.

The concept of 'liquefaction' was first patented in the US in 1914, but even though the science had been proven at the beginning of the century, the LNG industry did fully not take off until decades later – probably because there just wasn't enough demand for natural gas. By the time natural gas was more desired as a useful fuel in the 1970s, someone had fortunately already figured out how to transport it as a liquid. But it wasn't an engineer or a scientist who found the solution – instead, the answer came from a surprisingly left-field place: the meat business. In the 1950s, an American called William Wood Prince was the president of the world's largest livestock market, based in Chicago. This gigantic market slaughtered, processed, packaged and distributed meat across the continental US, and all that meat needed refrigeration, which meant electricity consumption. Wood Prince was struggling to maintain good profits with rising electricity prices, when he came across a fifty-year-old patent that sparked an idea.[3]

Wood Prince began studying ways to bring natural gas into Chicago, as an alternative fuel to power his meat refrigerators. He realised that if he liquefied gas in Louisiana (where it is abundant), it could be easily moved by barge up the Mississippi River, and then turned back to a gas on the other side. He could then use the gas to generate electricity at a potentially more stable price. Not only that, but Wood Prince realised that the process of warming up the gas releases waste cold – an extra and free source of refrigeration to chill the meat.

Teaming up with others who were looking to move natural gas, including the British Gas Council, Wood Prince converted the *Normarti* (an old Second World War cargo ship) into an LNG ship, renaming her the *Methane Pioneer*. She set off on her first voyage in January 1959, and moved 5,000 cubic metres of LNG – equivalent to about two Olympic-sized swimming pools full of liquid – from Lake Charles, Louisiana, in the Gulf of Mexico, to Canvey Island on the River Thames in the UK, nearly 8,000 kilometres across the Atlantic. The trip took twenty-seven days and marked the first commercial shipment of LNG.

Unfortunately for Wood Prince, after sparking the birth of a new industry, his plans to use LNG for his own purposes in the meat industry were scuppered when the FDA (the US government's Food and Drug Administration) were too worried about food contamination to approve the idea of cooling down meat with the cold generated from warming up LNG.[4] Wood Prince withdrew from the venture, presumably because the costs of using LNG did not stack up without using the waste cold. The *Methane Pioneer*, meanwhile, lived up to her name and pioneered the way for purpose-built commercial LNG ships.

A major natural gas discovery in the mid-1950s in Algeria kicked off a quest for fossil fuels and prompted the building of more LNG ships. In the early 1960s, the UK built two new ships to export gas from Algeria to the British Isles, the *Methane Princess* and *Methane Progress*, each with five times the capacity of the *Methane Pioneer*. These ships would feed Britain's demand for natural gas as the country switched from town gas. A liquefaction plant started up at the Algerian city of Arzew in 1964 to turn the natural gas to liquid and allow shipping of LNG to the UK and France, and eventually other countries. The naming of the ships did get more creative; the French ship *Cinderella* was eventually joined by *Prince Charming* to keep her company. Throughout the 1960s and 1970s, new liquefaction plants popped up in other gas-rich places, including Alaska, Libya, Brunei, Abu Dhabi and Indonesia. And the countries keen to receive the imported gas, like Japan, the US, Italy and later Belgium, Spain, Taiwan and South Korea, developed regasification plants to turn the liquid back to gas on arrival.[5]

As of 2021, there are more than 600 ships moving natural gas across the world.[6] When burned, natural gas emits less carbon dioxide per unit of energy generated compared to other fossil fuels like coal. While it is far from being emissions-free, it is the lesser of two evils. Turning it to a liquid and making the technology work at huge scales was a stroke of genius, and a testament to our ability to bring to life seemingly impossible solutions. It is proof that we can adapt and will be able to design new ways to move low-carbon energy sources across the world.

*

While I have spent some time on oil- and gas-industry ships and structures in the sea, my work has never taken me to a liquid-gas-carrying ship, unlike Captain Tatjana Pletena, who has been at the helm of the LNG ship *BW Everett* for the last few years.[7] I wanted to speak to Tatjana about her world, a tiny speck in the ocean, but getting hold of the captain was easier said than done. She was on the move constantly, and as a result her time zone shifted by an hour every other day. When we eventually managed to arrange a video call, she was somewhere in the South China sea.

Tatjana has overall responsibility for the ship, its cargo, and the thirty or so crew members. She described herself as an 'octopus' – with tentacles in all nooks of the ship to keep track of what is going on. Becoming an LNG ship's captain is an unusual career choice, to say the least, even more so for women, who make up just 2% of the seafaring workforce worldwide.[8] Tatjana told me that, at a younger age, she came across a picture of a girl in a naval uniform and white gloves in a careers book, and immediately thought, 'She looks really cool!' Without any knowledge or connection to shipping, she decided to go for a job at sea. Perhaps a superficial reason to choose a career, but it exposed her to a sector she thrives in at a much deeper level.

Tatjana graduated from the Latvian Maritime Academy in 2001 and started her career as a deck cadet with a Norwegian shipping company. The beginning was a shock; she had to adapt to a new way of life. Recalling the overwhelming smell of a gas terminal dominating those early days, she told me, 'It's like you're smelling the metal all the time, but you get used to it.'

I wondered what kept her at sea – what kept any of them at sea – away from friends and family for weeks and months at a time. Her original plan was to do the job for a maximum of five years, but she surprised me by describing the satisfying sense of freedom she feels at sea. This was the opposite of what I had imagined – the claustrophobia of being stuck on a vessel in a blue desert. Over the years, she had built up experience and risen through the ranks, becoming a local celebrity on her first day as captain. The crew were excited to be sailing in a fleet with a female captain for the first time, and asked to take pictures with her for their mothers, girlfriends and wives – a reassuring story for me to hear as a woman in a male-dominated industry.

Life aboard the *BW Everett* mirrors the global nature of LNG shipping, and energy in general. The crew is a mix of nationalities: Filipino, Chinese and Indian, amongst others. During the short periods of free time on the ship, they watch films, sing karaoke, play basketball or swim in the ship's pool when it is warm enough outside. The crew can change for every trip, so there is an initial period of trust-building and getting to know and understand each other's cultures. This is also reflected in the food; chefs are trained to cook for a multinational crew. Tatjana was pleasantly surprised to find the Filipino cook had gone to the effort of making borscht, an Eastern European beetroot soup, to make her feel more at home. All of these activities help to create a sense of comfort and normality in this unusual setting.

The ship runs on various fuel types, including the natural gas it carries. On the deck of the ship, mint green storage containers called 'membrane tanks' hold the LNG. There is enough space to carry 138,000 cubic metres worth of fuel (going back to Olympic swimming pools, that's fifty-five of them), a massive increase on the operation of the *Methane Princess* and *Methane Progress*. In terms of energy supplied, one shipload is enough to heat nearly 80,000 average-sized German or UK homes for a year.[9, 10]

Four box-shaped tanks seal the liquefied gas inside, making sure it does not leak into the ship's hull space and protecting the steel hull from getting damaged by the extreme temperatures of the liquid at minus 163 degrees Celsius. The boxes are insulated, keeping the LNG in a liquid state. As the gas warms up over time, a little bit boils off (under a quarter of a per cent per day). The 'boil-off' is diverted to the ship's boilers and used to generate steam from water, which spins the turbines that move the ship. It's a clever use of the valuable product, and the continuous evaporation takes heat away from the liquid, keeping the rest of the LNG cold, a process called auto-refrigeration.

When I spoke to Tatjana, she and the crew were dropping off some natural gas in China, but they did not know where their next destination would be. Their movement around the globe is dictated by the natural gas 'spot market', a global market for buying and selling natural gas. The vessel's company constantly monitors the gas market and gas prices, tracking who needs LNG moved from A to B, and calculating if it will be profitable to send the ship there. As such, the ship may unload all the gas at a port and travel to the next location to pick up more, without any boil-off to use as fuel. Because of this, the ship also carries other hydrocarbon fuels used for propulsion, instead of or in combination with the LNG boil-off.

Tatjana explained to me that, in some parts of the world, she has to use a mixture of fuels because of local requirements and emissions rules, as different fuels emit different amounts and types of pollutants when combusted. She could also be instructed to use a certain type of fuel over another, depending on the day's fuel prices. On the journey Tatjana was making when we spoke, they stopped in Singapore to refuel, in order to be ready for a potential thirty-eight-day voyage from China to Nigeria without any LNG 'heel' on board – a small amount of LNG usually left on the ship after delivering its load, used for propulsion and to keep the tanks cold, ready to be reloaded.

For an outsider looking in, being the captain of an LNG ship appears to be a risky job. It is worlds away from a desk-based nine-to-five and the stakes are much higher, yet despite this, Tatjana seemed relaxed. She clearly has a deep understanding of the risks, and explained to me that safety rules and requirements are very stringent. While the ship does carry a flammable gas, it is not going to ignite in its liquid state. It is a different story, however, if the liquid leaks from the tanks, turns into a vapour and combines with air to create a flammable mixture. 'We have detectors sensing the atmosphere all the time; we get a signal if there are any small leaks and deal with the situation,' Tatjana assured me.

One other safety issue was playing on my mind: what about the risk of pirate attacks? Tatjana agreed that it is a concern, especially around Africa. However, areas like the Gulf of Aden, between Somalia and Yemen, have improved since military vessels started patrolling the area. She recalled going to Nigeria in 2019 with a military escort and passing a vessel that had been attacked just an hour or so earlier. *BW Everett* has not experienced any attacks and has defences in place. For a start, it can travel fast, outrunning attackers. The main deck is high above the water, making an attack more challenging, and razor wire protects any areas accessible closer to the water. The high risks associated with this job are not for everyone, but I would feel safe aboard a ship with Tatjana at the helm. I can imagine her dealing with anything the sea throws her way.

Ships like the *BW Everett* are an energy bridge between producers and users, but shipping LNG is just one of many energy bridges provided by the shipping industry. Giant ships also move crude oil from oil fields to refineries, where it is turned into products like petrol for cars, jet fuel for aircraft, or heating oil for furnaces. The *Oceania*, built in South Korea in 2002, was the largest ship ever built at the time. Her parts arrived from different corners of the world, and thousands of people put the pieces together. Giant yellow cranes towered over the shipyard, lifting and placing the different sections in the correct positions; an eighteen-month-long, high-stakes and complicated Lego build.

We have gone to the extremes of constructing supersized ships to move energy over the vast distances that separate producers and consumers because our modern lives depend so heavily on large energy supplies, and because of the huge profits the industry gains. The ecosystem of extracting natural gas, turning it in to a liquid, shipping it and turning it back in to a gas has developed into a booming industry over the past sixty years or so. Until there is a radical shift away from fossil fuels, this is set to continue. My former drilling colleagues will continue to drill holes in the Algerian desert and elsewhere, in search of more gas to satisfy the seemingly ever-growing demand. And Tatjana and her team will continue to move LNG from where it is sourced to where it is needed.

When we do reach the point of a major transition away from fossil fuels, I like to think that LNG ships could be modified to carry other, more sustainable fuels. The first LNG ship was a repurposed Second World War cargo ship, after all. Crews can also transition, using their existing know-how and retraining. For the time being, though, LNG remains a major fuel for several huge energy-consuming nations – in 2022, natural gas met a quarter of our energy needs across the world.[11]

As nations look to move away from coal and oil, natural gas will be a tempting option, as it is readily available and emits less carbon dioxide than other fossil fuels. Making incremental and imperfect improvements is, in some cases, the best option. But in this case, I wonder if we are falling into the same trap that inhibited wind and solar power in the past, choosing readily available fossil fuels and delaying full decarbonisation of the energy system? While natural gas is abundant today, it will one day run out, so why not skip a step ahead and leave it in the ground?

As well as causing devastating climate change, this ongoing reliance on fossil fuels comes with other risks in the form of damage from fossil fuel spills into marine environments, or terrible loss of human life when things go wrong during transport by rail, road or sea. While LNG shipping has an excellent safety record, other means of moving energy come with greater risks. How far are we willing to go, and what damage are we willing to inflict in order to move energy around the planet?

*

Friday evening into the very early hours of Saturday 6 July 2013 was a sticky, hot night in the town of Lac-Mégantic, Quebec, Canada. Little did the inhabitants know, but the town was about to dramatically change forever. In the dead of the night, while the 5,800 or so residents slept in their beds, an unattended eastbound train started to roll down the tracks. It had been parked up for a crew change when a combination of factors caused the brakes to fail. As it rolled downhill, it accelerated to more than 100 kilometres per hour. The train hurtled on, eventually coming off the tracks and derailing in the town at around 1am. No one was aboard the train, but this was no ordinary train – it had started its journey in North Dakota, picking up 7.7 million litres of crude oil, bound for an oil refinery in New Brunswick on the east coast of Canada.

Events unfolded quickly. Once the first tank car derailed, others fell like dominoes. About six million litres of crude oil flooded out of the damaged metal tank cars, almost immediately going up in flames and causing explosions. Confusion followed; it took the train operator company some time to piece together what had happened. They issued a press release later in the day saying, 'At this time, we don't know how many cars are derailed ... We have reports of explosions and buildings in the city on fire, and a number of fatalities and injuries.'[12]

A total of sixty-three out of the seventy-two train tank cars had come off the tracks. The burning inferno of crude oil, combined with the pile-up of metal train tank cars, made the firefighters' job extremely difficult. Despite this, they did everything they could to limit the damage, but they couldn't save the forty-seven people killed in the blaze and explosions.[13] Two thousand people were forced to leave their homes, and the town was destroyed. The scene was so close to a Hollywood depiction of disaster that real footage from the tragedy was used in the popular Netflix film *Bird Box* (though the footage was later removed).[14]

As with any major disaster, an investigation followed to figure out what went wrong, what caused it and what lessons could be learned. Transport Canada issued an emergency directive to ban trains transporting dangerous goods from operating with single-person crews, and rewrote some of the rail operating rules. The investigation also led to changes in how unattended trains are secured, how liquid fossil fuel crude oil is classified, and how employees are trained. While this does not change the past or bring back the lost lives, it does help to improve the future.

Over the years, many lives have been claimed by the transport of fuel – these are the lengths we are prepared to go to, as a society, to move fuels around. However, each disaster highlights failures of the system, and prompts changes to be made to stop it happening again. While the modifications made following the Lac-Mégantic accident improved safety standards of fuel-carrying trains, the memory of the incident is understandably etched into the collective mind. Moving fuels brings benefit for those receiving the goods, but it comes with inherent risk.

The safety of the movement of fuels by engine-powered rail has come a very long way since its beginnings in the 1800s, spurred on by the expiry of James Watson's steam engine patent, which opened up the field of engine design to new ideas.[15] Compact versions of steam engines were invented, perfect for propelling a train along a track. Train networks took off in Britain, supporting the expanding coal-mining industry in two ways – firstly by moving coal from the mines to consumers, but also by using coal to power the engines that move the train itself. More trains meant more coal could be mined and moved, which in turn meant that more coal was needed to power the trains in an environmentally toxic cycle. The British population's demand for coal, used for cooking and heating well before the switch to town gas and subsequently natural gas, was insatiable, growing from five million users in 1700 to twelve million by 1831.[16]

Railways were not confined to Britain, of course – they began to sprawl across the US in the 1800s, moving coal as well as crude oil from oil fields to refineries.[17] Early versions of railroad tank cars looked very different to those that destroyed the town of Lac-Mégantic. They were wooden barrels, inspired by whiskey barrels, which sat on flat platforms on wheels. Shortly after the first iteration, bigger wooden barrels flipped onto their sides were used. Wood gave way to metal tanks, doing away with oil leaks from wooden barrels.[18] The method for oil-carrying rail and road tankers remains similar today – a horizontal metal tube filled with the liquid, mounted on a wheeled platform of some sort.

Oil-carrying trains actually play a small role in the movement of energy resources, relatively speaking. Before 2011, only about 0.2% of crude oil arrived at US refineries by rail. By 2020, this rocketed up to 2% because of increased domestic oil refining in the US – refining their own products meant fewer imports were needed.[19] This called for extra crude oil to be brought in, but there wasn't quite enough space in the pipelines that move the bulk of crude oil to US refineries. While the amounts coming in by train are tiny, any rise will understandably worry a community that has seen the impact of rail accidents.

As well as trains, many oil companies rely on trucks to move their product around. Oil trucks travelling along roads move an even smaller amount of crude oil to refineries than oil trains, but they also move some of the refined oil products to their final destinations. It's a convenient mode of transport, as trucks can go anywhere as long as there is a road, but it also introduces the risk of traffic accidents and oil spills. Driving an oil truck is a high-stakes game; the vehicle has a high centre of gravity, and contains heavy liquids sloshing around. The trucks can tip over more easily than other vehicles, so oil spills from trucks are more common than from ships or trains, but the volumes spilt tend to be much smaller, and so don't often result in tragedies on the scale of Lac-Mégantic.

That's not to say that it isn't without its dangers. Dashcam footage released by Michigan police shows a white fuel tanker driving along the highway in the city of Troy in July 2021, moments before it collides with the concrete barrier on the left-hand side of the road and tips over onto its side. The vehicle is engulfed in white smoke, before bursting into bright orange flames mixed with black smoke. Miraculously, the driver got out without being seriously hurt.[20] This dramatic scene immediately evokes an emotional reaction and the thought that we should not be moving flammable liquids by road like this. But, as with every aspect of the energy industry, we need to look at the reality of what this would mean for millions of people, especially those in more difficult-to-access areas.

Not having access to fuel would leave some people stranded in extreme deadly cold, as some remote homes use propane or furnace oil for heating, which are delivered by road. People would not be able to fill up their cars to drive to buy food or get emergency healthcare. For these people, the arrival of the fuel-filled cylinders on wheels is good news – in Hamilton, in New Zealand's North Island, the first ever petrol truck delivery in 1927 was paraded by a brass band.[21] While not all trucks are met with this level of enthusiasm, they are often welcomed by remote communities. For the time being, for better or worse, these trucks are an essential part of modern life.

People are opposed to the continual use of ships, trains and trucks to move energy resources for very valid reasons like environmental concerns and safety fears. So perhaps one obvious question is: why not replace all these transport systems with pipelines?

Chapter 6

Pipelines

AS A GENERAL rule, if I walk past any construction works, I will slow down to gawk at something that most pedestrians barely notice. In the spring of 2019, it took me longer than usual to walk anywhere in London because of the holes that began to appear in the ground all over the city. The cause of my distraction was a huge nationwide project to replace the UK's old natural gas-carrying underground pipes with shiny, modern plastic ones. Without any professional involvement in this project, these holes were my only window into the work. I would grind to a halt and curiously stick my face as close as possible to the metal mesh fence, to see if I could spot an exposed ageing iron pipe, grimacing at the state of it – dirty, crumpled metal flaking away. At street level, I'd also be on the lookout for neatly stacked bright yellow plastic pipe sections, waiting patiently to replace the crumbling old iron ones.

As well as exciting my nerd senses, these new pipes play a role in reducing greenhouse gas emissions, both immediately and potentially in the future too. Decaying iron pipes leak natural gas – methane – into the atmosphere, a substance almost thirty times more powerful at warming up the

Earth than carbon dioxide, so these yellow plastic pipes will immediately reduce the leakage and prevent further damage. Their future role may also be to carry hydrogen – a potentially low carbon alternative to natural gas – or some other blend of more environmentally friendly gas to fuel the UK's needs.

Natural gas may have started life somewhere far afield, like deep under the frozen ground in Siberia, but it travels many miles through a labyrinth of pipelines to get to my apartment, and eventually to my gas cooker or water boiler. It hasn't always been like this, and not everywhere in the world has gas on tap. When I was growing up in Baghdad, we had gas (which was probably a mixture of butane and propane, rather than the main component of natural gas, methane) delivered in small canisters, which would be hooked up to our cooker. But many of us are accustomed to an invisible, reliable supply. The arduous journey the gas has made to reach us when we flick a switch to turn on a heating system or use hot water does not get even a moment of thought. It's only when the supply is disrupted, or prices skyrocket, that the vital importance of pipelines becomes apparent.

Whether we realise it or not, for most of us, life is invisibly serviced by the pipelines that hide below the ground. Aside from the important job of moving energy resources, pipelines bring clean water into homes and carry away harmful sewage to water treatment facilities. Pipelines have been around for thousands of years – moving fluids in hollowed-out logs, or clay or bamboo pipes. It is thought that in ancient China, oil wells were drilled over 200 metres deep into the ground as early as the year 350 CE, using strong bamboo as the drill bit, and bamboo pipes to move the oil around. The fuel delivered by the pipes was burned for heat. In a stroke of genius from the ancient world, this heat was also used to boil seawater to separate out the salt and make it drinkable – a remarkably effective instance of early desalination.[1]

Pipelines took a little longer to take hold in the Americas; in the early days of oil in the US, horses clip-clopped along, pulling wagons loaded with oil-filled whiskey barrels, and barges cruised along waterways to move the same barrels over even larger distances. But progress was fast in nineteenth-century America, and by 1879 the Tidewater Pipe Company pushed pipeline engineering to the limit and built what was, at the time, the longest oil pipeline in the world: a 180-kilometre line across Pennsylvania from Coryville to Williamsport.[2] This kicked off a pipeline frenzy, and by the 1920s pipes criss-crossed the country, further spurred on by the growth of the motor vehicle industry and the rising demand for oil. The energy demands of the Second World War called for fast innovation too, and the US experienced a boom in pipeline construction both during and after the war, resulting in longer-distance, larger-diameter, and higher-pressure pipelines.[3]

It is difficult to overstate the scale of the energy interconnectedness brought about by pipelines; the US alone has just under four million kilometres of oil and natural gas pipelines, enough to wrap around the globe one hundred times over.[4,5] This scale makes pipelines inescapable for most of us, and certainly even more so for me; they pop up in every engineering job I do.

While international borders and large bodies of water can and do bring political challenges, the technical barriers are surmountable for a pipeline. The 1,224-kilometre twin Nord Stream 1 pipelines are the longest undersea gas pipelines in the world, creating a pathway for moving natural gas from Russia to Germany via the Baltic Sea – or at least they did, until the 2022 Russia–Ukraine war forced Germany to reassess its relationship with and reliance on energy from Russia. Prior to this, the two pipelines had moved gas into Germany ever since they started operating back in 2011 and 2012 respectively. Together, they could move 55 billion cubic metres of gas per year, just over half the amount of natural gas consumed by Germans in 2019.[6]

Constructing pipes like this is a complicated and delicate engineering achievement; imagine all the things you have to think about to build an underground pipeline – figuring out how big it should be, making the pipe itself, and making sure it is safe. Now put all that under the sea, and the complexity only increases.

Long before any Nord Stream 1 pipe touched the seabed, the pipeline route was mapped – a task not quite as simple as choosing the shortest path. A survey ship meticulously gathered data to create a three-dimensional model of the seabed, identifying every detail – any steep slopes, rock outcrops, environmentally sensitive areas and existing infrastructure were mapped. The pipeline takes up a miniscule surface area – it is like laying drinking straws across a football field – but there were still existing shipping lanes, fishing areas and protected conservation zones to avoid. Other obstacles included munitions: mines laid in the Baltic Sea, German weapons confiscated by the Allies at the end of the world wars that had been dumped in the area, and chemical weapons from the Eastern Bloc. The survey found 432 of up to 150,000 mines thought to be in the Baltic Sea. Any in the way of the pipeline route were carefully removed. It was not all weapons, though; the Baltic Sea is also a treasure trove of history. The low levels of oxygen and salinity and the absence of boring worms – worms that burrow into submerged wood in salty waters – make it an ideal environment for preserving shipwrecks. About 160 shipwrecks were found, sixty-four of which were thought to be culturally significant. The survey also listed paint cans, shopping trolleys, cars, washing machines and refrigerators, all sitting at the bottom of the sea.[7]

Once the route was selected, the pipe-laying ship, essentially a giant floating factory, came along to put the Nord Stream 1 pipelines in place. The crew on board worked 24/7 to weld twelve-metre sections of pipe together, to make one continuous long pipe, lowered off the back of the ship onto the seabed. Each pipe section was already coated in concrete ahead of arriving on the ship, to weigh it down and make sure the structure didn't float up off the seabed. Like squeezing icing from a piping bag onto a cake, the pipe came off the back of the ship in a 'lazy S' shape. The top of the S was supported by a 'stinger', a structure that held the pipe as it rolled off the back of the ship, and slightly resembled an insect's stinger. The bottom of the S was supported by its final resting place on the seabed. This left the middle section unsupported, and at risk of snapping, which was why the ship had to keep moving until the job was done.[8]

Undersea pipelines need some of the world's most unique experts and professionals. One such expert is Norwegian diver Einar Flaa. Looking at his pictures in media interviews, Flaa looks like a normal man with a normal job. But in fact, Flaa has one of the rarest and most specialised jobs in the world. At the time of construction of the Nord Stream 1 pipelines, he was one of only thirty or so technical divers in the world trained to weld pipe sections together at water depths of more than a hundred metres.[9]

I remember my amazement and slight jealousy when I discovered there was such a job as a technical diver. It was shortly after I began working in the oil and gas sector, and for a brief while I was quite captivated by the idea of becoming one of these rare divers. I imagined people's faces at parties when I told them about my job. But I had never dived before, and my romantic idea quickly dissolved after I visited a dive support vessel – the ship that divers live on when they are on duty. It was a large vessel, but the divers were confined to small, airtight chambers pressurised to match the depths they work at – roughly ten times the pressure we experience on land. This is to stop them suffering from the 'bends', or decompression sickness, where the rapid change in pressure causes potentially deadly gas bubbles to form in the blood.

Being one of these specialised divers is akin to being an astronaut. Before the internet, Flaa had to handwrite messages and pass them to staff outside the chamber he lived in, who scanned and sent them to his family in Norway.[10] For the safety of the divers, the whole place is monitored by cameras, watched day and night by the life-support technicians outside the pressurised chamber. Privacy does not exist (although Flaa and his colleagues were allowed to tape over the camera lens in the bathroom).

The divers working on the Nord Stream 1 pipeline construction were on the job for three weeks at a time, living in the pressurised chamber throughout, followed by three weeks off back at home. It took about five days at the beginning of the shift to allow the divers' bodies to adjust from atmospheric pressure to the pressure on the sea floor, and a similar amount of time at the end of the shift to adjust the other way. That left just over a week in the middle for actual work.

The pipe-laying ship crew completed the vast majority of the welding, laying the pipelines on the seabed as three long separate sections, which needed to be welded together underwater – a bit like melting the ends of three plastic straws together to make one longer straw. Building it in three separate sections saved on the amount of steel needed. This is because the three sections of pipe have the same internal diameter but different thicknesses, starting with the thickest pipe on the Russian side, where the pressure of the gas is highest at 220 bar – about 220 times higher than the pressure of the atmosphere pushing down on us. The pressure drops as the gas travels through the pipe, getting to about 200 bar midway, and ending at 100 bar by the time it reaches Germany. The thickness of a pipe depends on how much pressure it has to withstand. So, if all the pipe sections were identical, they would have to be thick enough to handle the highest pressure and use up more steel, hence this telescopic design.

Welding the three long pipe sections together was mostly automated, controlled by the *Skandi Arctic* dive support and construction vessel. A lot of the work was done by robots, but it would have been impossible without the human divers. 'Everything is run from up on the ship, but we're their hands,' Flaa said in a media interview.[11] Twelve divers lived in the sealed chambers on the *Skandi Arctic* vessel, and they took it in turns to descend to the bottom of the sea in a diving bell – a lift that carries the divers up and down between the vessel and the seabed. They went three at a time, for eight hours or so. Staying at a comfortable temperature was important, so the diving suits were warmed with hot water – a 'one-man sauna', according to Flaa.[12] It's a fine balance; the divers want to stay warm, but not too warm. If you start to sweat, you cannot wipe your brow because you have the suit on. If sweat pours down your face into your eyes, there's nothing you can do about it.

On the seabed, the divers had to feel their way around in the darkness, guided only by the light from the diving bell. They unpacked and mounted precision equipment for cutting the ends off the pipes. Large pipe-handling frames then lifted and shifted the pipeline segments, lining them up. At this point, a welding habitat was lowered over them. Water was pushed out of this box to create a dry zone. In here, the divers could work without their diving equipment on to set up the automatic welding machine. The closest thing I can think of to visualise this is the home of Sandy the squirrel from the cartoon *SpongeBob SquarePants*, who lives in a bubble-like dome on the seabed, filled with breathable air so that she can survive in there without her atmospheric diving suit on. Her sea-creature guests have to wear water-filled helmets in the bubble.

Inside the welding habitat, Flaa and the other divers had to line up the two pipe ends so they were completely flush. This had to be correct down to the millimetre, so they found themselves spending as long as six hours lining up the pipes. Once they were in the right position, an automatic welding machine finally joined them together. The completed weld was then inspected to make sure it was perfect, at which point all the equipment was lifted back up. The best part for Flaa was hearing the engineers say, 'It's a good weld, wrap it up,' through his earpiece.[13]

It took more than twenty million hours of work over the course of six years for the two Nord Stream 1 pipelines to start moving gas.[14] In 2021, the laborious construction of another set of twin pipelines – Nord Stream 2 – was finished. The intention was to double the capacity of natural gas that could be shifted from Russia to Germany via this route. While the pipes were technically ready to use, no gas flowed. As well as shutting down the original Nord Stream pipelines, Russia's invasion of Ukraine in February 2022 kept the new pipes idle. Adding to the political drama, on 26 September 2022, vast amounts of the natural gas that was sitting in the pipelines spewed into the Baltic Sea near Bornholm, Denmark, when several explosions damaged three of the four Nord Stream pipelines.[15] No one has claimed responsibility for the blasts, leading to a lot of international finger-pointing: some blame Russia, Russia has blamed the UK and US, and some are blaming a 'pro-Ukrainian group'.[16] Investigations are ongoing, but solving the mystery is unlikely to save the pipes; Germany and much of Europe have been forced to seriously reconsider their choice of energy partners.

*

While the technical challenges of these pipeline projects seem complicated and difficult, it is clear that they pale in comparison to the political and legal tangles. This is the reality of what it means to share the precious resources on our planet. Politics and energy pipelines in particular are inseparably intertwined.

Up until 2022, the European Union imported large amounts of natural gas from Russia; in 2021, Russia accounted for around 45% of the EU's gas imports.[17] Any fluctuation in gas flow from Russia to Europe sent shockwaves across the energy markets. In 2009, when Russia cut off gas supplies to Ukraine after disputes over price and accusations of political blackmail, prices in the UK shot up by as much as 27% in just one day.

Europe's past dependence on Russia and Russia's pre-existing international relations created some fierce opposition to the Nord Stream 2 pipeline project. The $11-billion pipeline was first announced in 2015, and while it was owned by Russia's state-backed energy giant Gazprom, heavy financial investment was provided by German, French, Dutch and Austrian energy companies. The involvement of these energy giants created an incentive for those outside of Russia to see the project succeed. Supporters argued that the pipeline would make Europe's energy supplies more secure, with a stable natural gas supply on tap, while critics saw it as handing Europe's energy security over to Russia on a silver platter – given the events of recent years, they had a point.

Zooming out to look at existing gas pipelines that were already moving fuel made the picture even more complicated. Alongside the Nord Stream 1 pipelines, there are other routes for Russian gas to reach Europe in overland pipelines, including ones passing through Ukraine. Nord Stream 2 created an alternative route, allowing Russia to bypass the overland route, depriving Ukraine of transit fees that made up 4% of the country's GDP.[18] This, added to Russia's repeated use of gas supplies as a political weapon against the country, justifiably set off alarm bells for Ukraine from the very beginning – they have fiercely opposed the pipeline since it was first proposed, just one year after Putin's first invasion of Ukraine in 2014. Poland, the Baltic states, Italy, UK and the European Commission were on Ukraine's side, pointing to Russia's actions in 2009 and threats to Moldova as evidence. And alongside the politics, there were market law problems too. European energy law required the owner of the pipeline to be different from the supplier of the gas flowing within the pipes, to prevent monopolies.

With decades-old rivalry between the US and Russia clearly a factor, the US joined this political battle, introducing sanctions on Nord Stream 2 in 2019. This paused the project for a year and a half, when the Swiss-owned pipe-laying contractor backed out. In 2018, Donald Trump tackled the subject at a NATO summit, saying, 'Germany will have almost 70% of their country controlled by Russia with natural gas. You tell me, is that appropriate? We're supposed to be guarding against Russia, and Germany goes out and pays billions and billions of dollars a year to Russia.'[19] On the other hand, the US has become an exporter of natural gas, so opposition to Nord Stream 2 could have been less than purely altruistic and motivated by the fact that they wanted a slice of the pie too. It is, after all, in their interest if Europe purchases natural gas from the US, shipped to Europe as LNG, instead of relying on Russian gas coming in by pipeline. This is what has started happening in recent years, as imports from the US and others have plugged the gap left in Europe by the lack of Russian supplies.

The Biden administration softened its stance on the Nord Stream 2 project, and the US sanctions were eventually lifted. In July 2021, Joe Biden and the former German Chancellor Angela Merkel agreed that the US may trigger sanctions if Russia ended up using energy as a political weapon – a measure that was deployed in response to the 2022 invasion of Ukraine. The deal aimed to stop Russia from cutting off gas supplies to Poland and Ukraine, but clearly did not quite achieve its aim. At the time, Ukrainian president Volodymyr Zelenskyy said he felt as if he had been stabbed in the back.[20]

The Nord Stream pipelines are now perpetually stuck between a rock and a hard place. The money has been spent and the physical infrastructure is sitting there; it is a huge waste to abandon the brand-new pipeline at the bottom of the sea. But it's difficult to imagine an alternative solution given the political situation surrounding it. Looking further ahead, with enough desire to move away from fossil fuels, the gas pipelines could become obsolete rather quickly unless they are adapted to move a future fuel like hydrogen – politics allowing, of course.

The Nord Stream nightmare is a perfect example of why energy systems cannot be viewed in isolated pockets. At the ends of the tentacles of these systems are people like you and me, parents playing with their children, elderly people enjoying time with lifelong friends, trying to get on with their lives. The severe impacts are felt in daily life – in countries where energy prices have risen to unaffordable levels. Worse still, there are people in conflict zones dealing with hugely disruptive power cuts and energy shortages.

The politics of pipelines stretch across time and space, changing the lives of the people in their way. While the Middle East may seem distant from Russia and Europe, the political ripples stretch to this region too. The war in Syria – which entered its twelfth year in 2024 – is fuelled by a mixture of factors interacting in a complicated way, far beyond my full understanding. But some experts have commented that energy pipelines play a part.[21] It could be argued that, to control the flow of energy, Saudi Arabia, Qatar and Turkey have a motive to remove Syria's president Bashar al-Assad from power so that they can run a pipeline through Syria, to Turkey, and into the European market. This would take a share of Russia's energy market, which could explain Syria's friendly relationship with Russia and its state-owned natural gas company Gazprom. Qatar, a huge exporter of LNG, approached Assad to propose the construction of a natural gas pipeline from Qatar through Saudi Arabia, Jordan and Syria to Turkey. In 2009, Assad refused to sign the agreement 'to protect the interests of our Russian allies'.[22] While energy distribution options may not have led directly to these conflicts, it is hard to believe that they did not play any role at all. In my own country of birth, Iraq's pipeline network originally extended to Turkey, Syria, Lebanon and Saudi Arabia.[23] But most lines have either been destroyed or shut down because of wars, conflicts and sabotage – too many of them over the years to detail. In 2023, the only international crude oil pipelines operating in Iraq linked the northern Kurdish region to Turkey's Mediterranean coast.

These stories repeat in various forms across all continents. In South America, Venezuela has oil and vast offshore gas reserves, but the political state makes it difficult to do business internationally; and in Colombia in 2018, the last remaining Marxist insurgent group, the National Liberation Army, claimed responsibility for the bombing of the Caño Limón pipeline. In South Asia, a proposed 'peace pipeline' stretching 1,724 miles across Iran, Pakistan and India never materialised because of strains between India and Pakistan. Similarly, the central Asian countries of Kazakhstan, Turkmenistan, Uzbekistan, Kyrgyzstan and Tajikistan have become oil and gas producers, but they are all landlocked and need pipelines to move their energy resources to the world market. The proposed Trans-Caspian Gas Pipeline project would have moved natural gas from Turkmenistan and Kazakhstan to Europe, via Azerbaijan and Georgia, bypassing Russia and Iran. The pipeline has been talked about for decades but is yet to get off the ground, facing opposition from Russia and Iran as well as economic challenges.[24]

Planning and building a pipeline to move energy is a delicate operation, with many moving parts. The physical engineering of a pipeline is challenging enough, but at least engineers and other professionals work together towards a common goal. The active conflict zones, fierce competition and tense political relationships between nations is a completely different brand of challenge, with leaders struggling for power and continued historical disputes playing out on the global stage complicating the movement of energy resources and causing ripple effects for individuals and populations far and wide.

*

Joe Biden's first day as the 46th President of the United States was a day of celebration for many environmental activists, indigenous communities and farmers. On 20 January 2021, Biden put the final nail in the coffin of a long-disputed and highly controversial oil-carrying pipeline, Keystone XL.[25] First proposed in 2008, the vision was to move oil from Alberta, Canada, over to American oil refineries. But this pipeline would not be like others: the oil it would carry is different to most of the crude oil that flows through various pipelines across the US, which is part of the reason for the extreme controversy.

Known as 'tar sands' or 'oil sands', this source of oil is a mixture of sand, clay, minerals, water and bitumen – and results in a very dense and thick form of petroleum. As the 'tar' in its name would suggest, at room temperature, it has the consistency of cold molasses or treacle. If you have ever tried to measure out a spoonful of molasses or treacle, you will know that it does not flow easily, a reason this type of oil is difficult to extract and move.

This difficulty has not stopped the industry from developing, though, and the process of extracting tar sands often involves injecting steam down a drilled well into the accumulation of oil. Heating the sticky oil, just like heating molasses, makes it flow out more easily. This 'unconventional' oil, as it is known, can be mixed with more conventional hydrocarbons to dilute it, so that it can flow more easily along a pipe.

With all the media coverage of the Keystone XL pipeline over recent years, it's easy to forget that a Keystone pipeline actually already exists and has been moving tar sands oil since 2010. It starts in Hardisty, Alberta, travels east through the Canadian provinces of Saskatchewan and Manitoba, before taking a sharp turn south across the border into North Dakota, South Dakota, and on to Steele City in Nebraska. The objective of the new pipeline, Keystone XL, was to increase capacity, and create a shortcut between Hardisty and Steele City – it would have formed the third side of a triangle to the existing pipe route. From Steele City, more pipes are already in place to move the oil onwards across the United States.

Opposition to the new pipeline came from all directions, and for all kinds of reasons. The proposed route covered nearly 2,000 kilometres, with the vast majority running through privately owned agricultural land and rangeland. Landowners were understandably reluctant to hand over sections of their land for a pipeline (highlighting one of the benefits of offshore pipelines to their investors – there are no people in the way).

Along the pipeline route in Nebraska, farmer and landowner Art Tanderup was delighted to hear Biden's decision. He had been offered $20,000 by the pipeline company to give the right of way for the pipeline to pass through his land, but had fought for ten years, standing alone as the neighbours around him caved in, taking the money and signing away parts of their land. Tanderup was concerned about oil leaks contaminating water – his land is right above the Ogallala Aquifer, an important source of fresh water in the US.[26]

Tanderup's water worries are not unfounded, and were shared by many others, spurred on by past oil spills from pipelines. The notorious 2010 Enbridge pipeline spill released tar sands oil into Talmadge Creek, a waterway that flows into Michigan's Kalamazoo River. The spill shone a spotlight directly on the difficulties of clearing up this type of heavy, sludgy oil – it sank to the bottom of the river and got mixed in with the fresh flowing water. The US Environmental Protection Agency ordered Enbridge to dredge the river to get rid of the oil, but of course the damage had already been done.[27] While one oil pipeline leak is not an indication of future leaks, the Enbridge spill did draw attention to Keystone XL and raised concerns about leaks into waterways and contamination of drinking water.

Another legitimate problem was around the greenhouse gas emissions. As tar sands are heavy hydrocarbons, with a higher carbon content and less hydrogen, more energy is needed to extract and to process them into usable fuels. The increase in energy needed translates into more greenhouse gas emissions. However, this sole project itself would not have meant a huge increase in emissions on a global scale; this was one of the arguments made by supporters of Keystone XL. Still, others argued that it would send the wrong message and contradict any climate promises. Alberta's tar sands reserve also happens to be under Canada's boreal forest, which has been described as the 'world's largest and most important carbon storehouse'.[28] Clearing sections of forest to make way for fossil fuel extraction, no matter how small, goes directly against the flow of tackling climate change.

It is too easy to look at the science and engineering and come to a simple and obvious solution – leave the sludge in the ground, with the forest above intact. But Keystone XL came to embody more than the sum of its parts – it became a battleground for people's futures. While many indigenous communities in the US and Canada felt vindicated by Biden's decision on Keystone XL and urged his government to take action on other planned pipelines,[29] Alvin Francis, chief of Nekaneet Cree First Nation in Saskatchewan, had a different view. As a part owner of Keystone XL with over 500 people living on his First Nation reserve, and more than half of them unemployed, Chief Francis had to make some difficult decisions. He could either stand by and watch the pipeline go through his traditional territory, or capitalise on it and create work opportunities for his young people. With the pipeline now cancelled, he has to find alternative options.[30]

Just as Chief Francis had pinned his hopes on the pipeline to support his community, Alberta's tar sands industry employed around 120,000 people in 2020.[31] Some of these individuals have few other opportunities to make a living, and the tar sands provide a lifeline and a community for them. This is not a defence of the industry or a good enough reason to make the wrong environmental decision, with potentially devastating impacts on millions of people in other parts of the world. More so, it is a reminder that energy transition involves people, whose livelihoods are tied up in the current system, and that no one should be left behind. Chief Alvin Francis should not have been put in a position where the only hope for his community is an oil pipeline.

The Keystone XL pipeline has been laid to rest – at least for now. Much like other major pipelines, this one is beholden to the whims of politics, and it's already been on a twisting and turning path. Canada's government supported Keystone XL all along, but in 2015 the US State Department, under President Obama, declined to grant the permit needed to build, maintain and operate the pipeline. That should have been the end of the story, but on his fourth day in office, in a dramatic turn of events, President Donald Trump signed an executive order to revoke Obama's decision and bring the zombie pipeline back to life. A barrage of lawsuits were thrown at him, stalling the project for his time in office, until the 2021 final Biden U-turn. It will remain up to governments, with pressure from populations, to make the right calls on building pipelines – weighing up who wins and who loses in each case.

*

These political pipeline sagas may seem irrelevant in a fossil fuel-free future. As our use of the greenhouse gas-emitting fuels diminishes over time, the need for pipelines should follow the same trend, which would be a tremendous waste of infrastructure, especially of newer pipelines that have a lot of life left in them. But there is an alternative role for pipelines to play in a world of renewable energy, and it brings me back to those yellow plastic pipes I saw being placed in holes in the ground in London. These brand-new gas distribution pipes could carry hydrogen instead of natural gas.

My urge to stick my face into mesh wire fences to feed my curiosity for hydrogen infrastructure finally subsided in 2022 when I started working in hydrogen (although I continue to peek into construction works on a very regular basis). I suddenly got a glorious access-all-areas industry pass, and now have a front-row seat for this potentially exciting future fuel. Just like natural gas, hydrogen, the first element in the periodic table and the most abundant one in the universe, can be burned to release heat for industrial processes or homes, consumed in a fuel cell to power vehicles, or used as an ingredient in the chemicals industry to make some of the products we use daily. When combusted, hydrogen reacts with oxygen to release only water – no carbon dioxide – immediately improving air quality and dramatically reducing greenhouse gas emissions.

But there is, of course, a catch. The most abundant element in the universe rarely exists alone on Earth – it attaches itself to other elements to form molecules. Water, for example, is made up of hydrogen and oxygen, and hydrocarbons are formed of hydrogen and carbon. So, before any hydrogen can be moved or used as a fuel, it has to be separated out, or 'produced'. Some underground pockets of hydrogen have been found in Mali and in France,[32] and others across the globe are now scouring the ground beneath them for this fuel, which may provide an alternative to the production of hydrogen in the coming years.

For the time being, the main method of producing hydrogen, called 'steam methane reforming', has been widely used to produce hydrogen for making chemicals like ammonia, mainly for fertiliser, since the 1920s. It involves mixing a hydrocarbon fuel like natural gas with hot steam, giving it enough energy to split apart. This releases the hydrogen, but it also releases carbon dioxide. At the moment, the carbon dioxide from hydrogen production for chemicals manufacture is all released into the atmosphere, adding to the greenhouse gas inventory. For a low-carbon hydrogen, the carbon dioxide has to be captured and stored – something many governments and energy companies are heavily investing in at the moment. This type of hydrogen is sometimes called 'blue hydrogen' (I have yet to meet anyone who can tell me why the colour blue was chosen).

The other commonly available method of producing hydrogen is electrolysis, which involves passing an electric current through water, splitting it apart into hydrogen and oxygen. If the electricity used to power the process is renewable, the resulting hydrogen is deemed low-carbon and can be called 'green hydrogen'. Given that we want to stop our reliance on fossil fuels, this seems a better way forward than the 'blue' version, although it does raise the question of where all that renewable electricity and water – a vital resource – will come from. Seawater is one option, but the salt has to be removed first, adding to the amount of energy needed to produce the hydrogen in the first place – yet another challenge for green hydrogen producers.

While hydrogen may seem like the perfect wonder-fuel, the processes to produce it are energy- and resource-intensive, so this valuable substance should be used sparingly, with a large focus on energy efficiency. With hindsight, fossil fuels should have been treated in this way too. Had we built all homes to be well insulated, then, over the years, lower quantities of fossil fuels would have been used up to heat or cool them. Introducing a new, clean fuel is an opportunity to start again and do things better.

If electricity is needed to produce hydrogen, why not just use the electricity directly instead of going through the rigmarole of converting it to hydrogen? Where possible, electrification does make sense, but it comes down to the infrastructure and available cables to move the electricity. At the moment in the UK, the gas networks can meet a peak in energy demand about four times higher than the capacity of the electricity network. So, on the coldest day of the year, the natural gas pipelines can deliver enough energy to keep everyone warm, whereas the electricity generators and cables may struggle – if everyone in the UK switched to electric heating tomorrow, their heating needs might not be met on the coldest days. While it is not yet clear what shape the future energy system will settle into, it seems as if hydrogen could be an important feature, working hand in hand with electricity.

Hydrogen can be produced from different sources in different places, as it is manufactured rather than existing as a raw material or energy source. Unlike fossil fuel extraction, it is not geographically tied down, so it should not in theory be as easily influenced by geopolitics.[33] However, given the vast quantities of renewable electricity and water needed to produce green hydrogen, it is possible that it would be produced in certain parts of the world with the right resources for the process, and then moved elsewhere, just as oil and gas are extracted and moved now. Moving away from fossil fuels will certainly help to clean up the environment, but it may not solve the political issues. Unequal distribution of natural resources means some places will be able to produce more low-carbon fuels, bringing in a new, if different, era of energy and pipeline politics with new players in control.

Sarah Kimpton, Vice President at the Norwegian engineering company DNV, has spent over thirty years in the natural gas industry, but more recently moved over to low-carbon fuels.[34] During an animated video call, she painted a picture of the future of pipelines for me. Even after a major transition, when it comes to politics, it is possible we will face 'the same old thing', Sarah warns. Europe is already developing a 'Hydrogen Backbone', a series of pipes to move hydrogen between European countries. As an island, the UK is looking into how undersea pipelines can be used to move hydrogen. Sarah says Russia is also looking into converting its natural gas into hydrogen and putting that into pipelines, so if the political situation ever allows, the Nord Stream pipelines may one day morph into something slightly different and more fitting with a more sustainable energy system.

While it sounds like a drastic change, Sarah told me that, ultimately, you are still moving molecules in a pipe, be it hydrogen or natural gas. One of the main differences comes from the lower energy density of hydrogen by volume – a balloon filled with hydrogen has less than a third of the energy content of a natural gas-filled one. On a more local level, your gas hob still needs the same amount of energy to work, so in practice it means hydrogen will need to flow faster down the pipe to give the same amount of energy as natural gas. 'The molecules are whizzing along a lot quicker,' Sarah summarised.

Engineers are looking into what this means for existing pipes – can they handle the extra whizzing? Initial tests look promising. For an individual in a home, residents will notice that a larger volume of gas is flowing past their meter, so it will be important to continue billing people based on the energy content of the fuel they use rather than a volume of gas. There are a lot of details to iron out, but I find it encouraging to think that all those new yellow plastic pipelines might not go to waste if and when we move away from natural gas.

Just as gas pipelines could transition to carrying a lower-carbon fuel, oil pipelines in the future could carry synthetic liquid fuels. The liquid fuels we use today, like oil, are chemicals made up of hydrogen and carbon. A synthetic liquid fuel could be made by replicating this chemical structure – so combining a low-carbon hydrogen, perhaps made by electrolysis, with carbon. The carbon could feasibly come from carbon dioxide captured from the atmosphere, or from the emissions of an industrial process like making steel or cement. These synthetic liquid hydrocarbons could replace petrol for cars, or jet fuel for airplanes. At the moment, a large portion of the demand for liquid oils is for transport, so if the sector moves to electric vehicles, the need for these synthetic liquid fuels and therefore the pipelines to move them is likely to be smaller than today.

All of these prospects are a chance to do things differently and to change how the endless tangle of energy pipelines across land and sea serve society. As the energy system shifts and morphs into something fit for the planet, different methods of moving fuels of the future – hydrogen, synthetic liquid fuels, or low-carbon electricity – will have to keep up with the changes and adapt, like a crustacean outgrowing its shell and making a new one.

Chapter 7

Cables

THROUGH THE LONG clear glass panels of the viewing platform, I looked down. Below me was a cavernous room filled with screens, attended to by twenty or so people quietly delivering flawless power. At eye level, directly ahead of me on a giant screen, was an uncooked spaghetti-like mess of lines, lit up with small circular green and red lights. I stared at it, perplexed for a moment, until I tilted my head about forty-five degrees to the left, and realised these pinpricks of light added up to a map of the UK, on its side.

I was in the control room of the National Grid Electricity System Operator (or just the National Grid for short), the neurological centre of England's electricity network. In here, engineers make sure electricity is flowing the length and breadth of the country, from the rolling hills of the Lake District to the bright lights of Piccadilly in London. The two-dimensional lines on the screen ahead of me represented the electricity transmission wires. These are the metal cables mounted on pylons, carrying electricity up and down your country, wherever you are in the world. The green and red lights on the map indicate the health status, helping the National Grid engineers in this control room to keep track of any faults.

I was almost at the end of my chemical engineering studies at this point, but I knew very little about life as an engineer in the real world. I arranged this day out to the National Grid control room during my three-month work placement at a gas-fired power station, to try and piece together the puzzle of the energy industry. It all clicked for me during that summer; my eyes were opened to the wonders of energy networks, and the role power stations play in all our lives. I suddenly understood the potential applications of my studies and the difference I could make for society as an engineer in energy – although I had yet to fully understand this energy system's historical impacts on the climate and environment.

Running a gas-fired power station, it turned out, was a complex team effort, meticulously planned and managed over the years, even down to the minute details of how, when, and what chemicals to use to wash the blades of the gas turbines to maintain peak performance. All of this effort generated electricity. But what happened to the electricity after it left the power station? How did it move; where did it go? And how did this work across the world?

I find it difficult to understand concepts I cannot see and touch, which makes electricity quite tricky. Electricity is the presence or flow of charge; the charged electrons between atoms, far too tiny to see. But we can visualise the small particles passing charge to one another down the chain, causing charge to flow like water along a pipe.

Voltage, measured in volts, is the potential to push electrons around an electrical circuit, causing them to flow. This results in a current, which is the flow of charged electrons. Without the push of voltage, the electrons would move around randomly. Voltage is comparable to pressure in a water pipe – a higher pressure means that more water will flow past a certain point at any moment in time. Electrons power electrical devices, such as a light bulb, by flowing through them. A higher 'pressure' or voltage means more electrons pass through the light bulb, making it shine more brightly. Too much voltage will blow the bulb, as too many electrons try to squeeze through at once. Too little voltage and a lightbulb will dim.

Electricity as it is used today has been around since the early 1800s. The phenomenon was known long before this, but more as a scientific curiosity than anything useful; it took people a while to figure out a practical use for it. In an age of darkness, electricity for lighting was a revolutionary application, and heralded the beginnings of the electricity grid – the name given to the system that moves energy from power station to the end users through cables, and the namesake of the National Grid.

The first electric grid was constructed in 1879 to create lighting for gold mines in the Sierra Nevada mountains of the US, doubling the time that miners could spend searching for gold. Electric lighting for the media closely followed, with the *San Francisco Chronicle* adopting an electric arc lamp system later that same year, while the *New York Times* wired its offices for light in 1882. Reading tiny print by electric light was a vast improvement over the flickering light produced by candles, gas flames or whale oil lamps.

As this life-changing technology captured the attention of scientists, inventors and those in business, it prompted a battle of minds like no other, which decided the fate of electricity around the world. The story of the movement of electricity is filled with passion, cruelty, lies and betrayal, and is dominated by three main characters: on one side, Thomas Edison, a familiar name synonymous with invention; and on the other, a duo of colleagues who stood together, American engineer George Westinghouse and Serbian-born inventor Nikola Tesla (whose name was adopted by the US company famous for their electric vehicles).

The big debate was about how electricity should be moved. Edison insisted that direct current was the best method, while Tesla and Westinghouse championed alternating current. The terms 'direct' and 'alternating' describe how electricity flows. Direct current, DC, pushes electrons in one direction, whereas alternating current, AC, causes electrons to alternate between flowing forwards and backwards. To use a water analogy again, DC is like a river, flowing in one direction, whereas AC is like waves crashing onto a beach, then washing away in the opposite direction.

Edison, an experimentalist and sharp businessman, had built up his skills from a young age through a string of jobs, including selling sweets and newspapers to rail passengers, and employing other boys to help increase his profits. By the time the war of the currents took place, he had established himself as an authority on invention and electricity; he was a celebrity and 'wizard' of sorts. He was determined to realise his vision of every city having a small power station for every square mile within it, distributing DC electricity at around 100 volts through underground cables. The grids could not physically extend beyond a mile, because of the limitations of DC only flowing in one direction, with the power weakening the further away it gets from the power station. In Edison's plan, long-distance transmission was not possible, as the technology to enable this did not exist at the time – although it does today.

Tesla and Westinghouse argued the case for moving electricity by AC. The voltage of this sort of system can be increased, or 'stepped up', and decreased, or 'stepped down', by using an electrical piece of equipment called a transformer. Passing the electricity through a transformer – which only works for alternating current – and stepping up the voltage means it can travel longer distances, in the same way that water pushed through a pipe with a higher pressure can travel further. At the time, the voltage of DC could not be stepped up and down like this, so it could only travel shorter distances. In the Tesla–Westinghouse vision, long-distance transmission was possible; one big power station on the edge of town could do the trick for a whole city.

The AC method relied on the genius of Tesla, who invented the alternating current motor in his mind at the age of twenty-six. Tesla had experienced visions and hallucinations from a young age, and could build fully functioning machines in his head, tinkering and improving things, and perfecting concepts before constructing anything in the real world. He knew AC was a better system, but it took him a long time and a lot of heartache to get someone to take his side. In 1884, he travelled to America to work for Edison. Tesla told the great inventor about the benefits of AC, but Edison brushed the idea aside and instead asked him to improve the efficiency of DC systems, promising him $50,000 – well over a million dollars in today's money – if he managed it. Tesla completed the task within a few months, significantly improving a system he knew was inferior to his own. When he asked Edison about the $50,000, Edison replied, 'Tesla, you don't understand our American humour. When you become a full-fledged American, you will appreciate an American joke.'[1] Tesla promptly resigned and eventually found his way into business with Westinghouse, who saw the power of what Tesla had created. Westinghouse paid a substantial amount of money for his AC invention patents and agreed to give him royalties for every unit of electrical capacity sold.

The war of the currents ended with alternating current as the clear winner, and AC remains the dominant method today. Ironically enough, both Westinghouse and Tesla were later awarded the Edison Medal, an award created by a group of Edison's friends and associates: Westinghouse was honoured for 'meritorious achievements in the development of the alternating current system', and Tesla for 'meritorious achievements in electrical science and art'.[2]

In modern power stations, AC electricity is generated at a high voltage, which is increased to over 100,000 volts on its way out of the power station through transformers. It enters transmission cables, just like the ones I saw represented on the giant screen of the National Grid's control room. Some of the electricity is lost from the cables as heat, but moving it at a higher voltage reduces the loss of energy along the way. On the flip side, high voltage, regardless of being AC or DC, is extremely dangerous, so high-voltage cables have to be carefully protected to lower the risk of deadly electrocution. From the high-voltage transmission cables, the electrons flow into the lower-voltage distribution network cables, and the voltage continues to be reduced until the electricity enters a home and sits at the socket, waiting for a phone charger or toaster to be plugged in.

While the concept of going from high to low voltage is universal, the voltage level when it enters our homes varies across the world. Most of the North American continent, for example – from Alaska to Yucatan – has sockets at 110 to 127 volts – but much of the rest of the world is at 220 to 240 volts. This seemingly random division goes back to the beginnings of the electricity system, when there was no standardisation, so engineers chose the voltage that they thought was best. Plugging in my UK hairdryer in Canada was a disappointing experience; it blew out a weak stream of hot air no stronger than a breeze on a summer's day. Not enough electrons were passing through my device to power it properly. The 110 volts in Canada did not quite cut it for my hairdryer, designed to work at 240 volts.

Westinghouse and Tesla gave us the blueprint for moving electricity, but each country went on to build its own flavour of it. Many have a national grid that is looked after by one or more companies or governments. In England at the moment, the private company National Grid Electricity System Operator has control of those major arteries of alternating current, which works well since it is a small island nation and it's relatively easy to transmit electricity around. But in bigger landmasses such as the US, a grid may only cover a portion of the country; for example, Pacific Gas and Electric Company operates transmission cables across California only. Some states are only very loosely connected to the grid of the next state over, making them vulnerable if there are power outages, as they can't rely on their neighbours to help them out, whereas others have better interconnectedness. All these slightly different systems, connected to a variety of people at the other end, come with diverse challenges across the world.

*

At 8:30 p.m. on Thursday 5 April 2001, twenty-two million people in the UK simultaneously switched on kettles, flicked on lights, and popped some toast into their toasters, straining the National Grid to its limits. They had just watched one of the most dramatic plot developments in the history of UK television, when the *EastEnders* character Lisa Shaw admitted to shooting her former boyfriend, Phil Mitchell. The British soap opera, which has been going strong since 1985, follows the unusually dramatic and convoluted stories of some fictional residents of the East End of London. Lisa's confession resulted in a huge surge in electricity demand of 2,290 megawatts, five times higher than the normal surge seen by the National Grid at the end of an average *EastEnders* episode.

This phenomenon is known as a 'TV pickup'. It only seems to occur in the UK, where the population sporadically unites to watch the same television programmes in vast numbers, apparently more so than other nations. The stereotypical British love for a cup of tea and the prevalence of electric kettles may also be to blame, as people simultaneously switch them on to boil water for tea (presumably to calm the nerves after a big cliff-hanger ending). Such a surge in energy demand creates a challenge for the National Grid, who have to do whatever they can, including turning on extra power stations, to generate more electricity in order to meet the demand. A similar surge in electricity use happened after the Royal Wedding of Prince William and Kate Middleton in 2011, the England versus Australia match at the 2003 Rugby World Cup, and the 1990 Football World Cup semi-final game between England and West Germany.

The 'TV pickup' may be unique to the UK, but surges in electricity use for other reasons, such as unexpectedly hot days causing everyone to turn on their air conditioning, are a common challenge anywhere with an electricity grid, creating a complex problem. Electricity cannot easily be stored, so it has to be used the moment it is produced, which means that electricity grid operators have to balance the supply with the demand on the system. They estimate how much energy will be needed over the next year, month, week, day, hour, minute and second, and buy the corresponding volumes of electricity in advance from energy suppliers, who commit their power stations to generate these amounts.

However, there is always a risk that a power station does not deliver the amount it committed to, or that something unexpected happens on the demand side, like a nerve-wracking penalty shootout. Electricity grid operators have some tricks up their sleeves to make final adjustments and make sure supply meets demand. If there is over-generation, they can force power stations to shut down. If there is under-generation, they can bring on some fast-acting electricity generators. While effective, these can be expensive last-minute tricks, as the electricity grid operator has to pay generators to make the quick changes.

Peak demand events, or times of year when electricity consumption is at its highest, are relatively rare – counted in hours per year, rather than days or weeks. Traditionally, these are dealt with by building flexible electricity generation that can be quickly turned on and off, such as gas-fired power stations much like the one I worked at. This flexible generation is there to be used during extreme events, remaining idly on standby for the rest of the year. The large cost of building these power stations, coupled with the last-minute requests for changes in electricity supply from generators, can drive electricity prices up a hundred-fold during a peak demand event.

Instead of increasing the generation of electricity to meet the high demand, the problem can also be attacked from the other side. As individuals, electricity users have a lot of collective power. According to one calculation, if one in ten people turn on their washing machines at the exact same moment, it would bring the electricity system to its knees.[3] But these users can be asked to help by reducing their consumption – this is called demand side response, or DSR. One of the flavours of DSR is known as 'behavioural', meaning that consumers are prompted to change their behaviour to reduce energy use.

Australia's state of Victoria tested this out on a large scale, starting in summer 2017, when residents took part in a giant experiment dubbed the 'Curb Your Power' programme. Over 10,000 Victoria residents signed up to be called on to curb their electricity use during peak demand events. The principle is very similar to 'loadshedding', experienced by residents of countries like South Africa, with a very important difference – there is no choice when it comes to loadshedding. In Victoria, text messages alerted customers of the situation, asking them to use less electricity for the next couple of hours.

Each customer was set a personalised reduction target, based on their historical use. If successful, they were rewarded with a AU$10 discount on their next electricity bill.[4, 5] The trial, which lasted for around three years, managed to provide the grid with an average 0.45kW demand reduction per customer: that's 4.5MW for 10,000 participants. If scaled up to a million people, this could reduce demand by 450MW, enough to provide real support for the grid during a peak demand event – such as a normal TV pickup in the UK.

Behavioural demand response, like Curb Your Power, eases the burden on the electricity grid by changing consumer behaviour. A large number of people reducing their electricity demand can make a significant difference, lowering the need for spare electricity generating capacity that remains on standby for times of peak demand. However, it does rely heavily on the consumers – you and me – changing our behaviour; getting direct feedback on our energy use has been shown to encourage this. If we can immediately see the impact of the changes we make, for example with a visual representation of the reduction in consumption when the air-conditioning temperature is turned up by a few degrees, it motivates us to keep going. Playing on basic human psychology can spur on behaviour change – for example, showing individuals how much energy they are using compared to their neighbours creates competition, as it becomes a game to see who can use the least. Demand reduction programmes can use gamification and rewards to incentivise electricity users to help the grid.

In contrast with other ways to reduce emissions, like replacing fossil-fuelled electricity generation with renewables, the

Curb Your Power experiment was unique in its reliance on people taking action. Arguably, this is one of the most environmentally effective measures – with the potential to shift energy demand to times when there are more renewables on the system, and removing the need to generate electricity at peak demand times, along with the emissions associated with it. But it relies on all of us changing our habits.

Demand side response has been applied across multiple countries in other forms, usually in commercial buildings or industry, where controls are installed onto pieces of equipment to automatically turn their electricity use up or down, within certain limits, to help the grid. This may be a frozen food warehouse, consuming electricity 24/7 to maintain the freezer at minus twenty-five degrees Celsius. During a peak demand event, the automatic controls could turn the temperature up to minus twenty degrees Celsius for a quarter of an hour. It makes no difference to the frozen food, but when combined with other types of demand reduction at the same time, it makes a huge difference to the electricity consumed and the operation of the grid.

For the time being, demand side response is mostly limited to industrial and commercial electricity users, but this is changing with the digital age. Victoria's Curb Your Power programme was only available to customers with smart meters, devices that measure electricity consumption more accurately and transmit the data to the electricity provider every thirty minutes, eliminating the need for manual readings. Smarter electricity systems, smart meters, and smart appliances are flinging open the doors to demand side reduction opportunities, and the environmental benefits they bring. The future smart home, kitted out with a smart fridge-freezer, smart washing machine and electric vehicle, will monitor the status of the main electricity grid. Is it a windy, sunny day with too much renewable generation? The washing machine and dishwasher can turn on to soak up the excess electricity, and the electric vehicle can charge. A sudden surge in electricity consumption? The smart fridge is allowed to get a few degrees warmer (up to a safe level for the food inside), reducing its need for electricity and the burden on the grid. The changes are small, but collectively, across millions of homes, add up to reductions in energy consumption and the associated emissions.

This vision will not work without interconnectedness between appliances, homes and national grids, and potentially multiple connected grids across different countries. The different parts of the system need to talk to each other to communicate what is happening – all the while keeping safety and data security as a top priority. As well as following agreed communications protocols, operating alternating current at approved voltage levels helps to connect us. But the system is still not perfectly uniform, and this disconnectedness can lead to major problems.

*

Tsukiji market in Tokyo was once the largest fish market in the world. It buzzed with activity; tuna carcasses whizzed through the aisles on motorised 'turret' carts, alongside the slower old-fashioned wooden carts. Expert fishmongers with sword-like knives sliced and diced the fish into chunks, to be sent to prestigious sushi restaurants across the world. That was, until 11 March 2011, when the main island of Japan was struck by the earthquake and tsunami that flooded the Fukushima Daiichi nuclear power plant, an important source of electricity for the area. Alongside Chernobyl, the accident remains one of the worst nuclear disasters in history.

Thirteen days on from the Fukushima Daiichi nuclear disaster, rolling blackouts continued to plague Tokyo. The loss of the nuclear power station left a huge hole in the region's electricity supply. The utility company serving the area, Tokyo Electric Power Co., lost a fifth of its generating capacity, forcing it to cut people off from their power supply. Train services – powered by electricity – were also reduced. Rail commuters struggled to get to work on time, and rather than eating out after work, they hopped on any available train to make it back home. Restaurants had fewer diners, dramatically affecting the sales of fish from Tsukiji market to restaurants. Activity in Tsukiji market wound down to almost nothing. One of the market's wholesalers blamed the rolling blackouts, and predicted that a third of the traders in the market would close their businesses.[6] The market is just one example of the far-reaching havoc that blackouts can cause to individuals and their livelihoods.

What is astonishing is that Japan did have spare electricity generation to help plug the gap left by the nuclear power station – but the problem was that they couldn't move it from where it was generated to where it was needed. Unlike the rest of the world, Japan's electrical grid is divided. Countries with large transmission systems usually operate at 60 Hertz in the Americas, and 50 Hertz elsewhere. The frequency, measured in Hertz, describes the number of times alternating current changes direction back and forth every second. For a connected electricity system to function, all power stations in that system need to work in synchrony, operating at the same frequency.

Japan, an archipelago, is made up of four main islands and thousands of smaller ones. Its electrical grid is divided into ten separate zones, as well as two areas that operate at different frequencies. While the eastern part of Honshu (home to the Fukushima power station, the Tokyo metropolitan area and its fish market, and the island of Hokkaido) uses 50 Hertz, the western part of Honshu (including Nagoya and Osaka as well as the islands of Kyushu, Shikoku and Okinawa) uses 60 Hertz.[7] Because of the difference in frequencies, the western part of Honshu, which was less affected by the disaster and still able to generate electricity, was tragically unable to transmit the electricity to where it was needed in the eastern part of the island. Millions of people, including the fish market, were left without power.

The frequency quirk, which was such a devastating barrier in the aftermath of the 2011 natural disaster, can be traced back to the 1800s. The Tokyo Electric Light company bought 50 Hertz equipment from AEG in Germany, and similar companies in Osaka bought 60 Hertz equipment from General Electric in the US. In the early days, there was very little agreement about the frequency that alternating current should alternate at. There were many options available, from as little as 25, up to 125 Hertz.[8] It probably didn't seem like a big deal at the time, but it has resulted in a country divided into two halves, unable to exchange electricity with each other. At the time of the disaster, three frequency conversion stations existed between the two zones, with the ability to convert and pass just 1,200MW in total from one side to the other. This was nowhere near enough; the Fukushima Daiichi nuclear power station had a generating capacity of about 5,000MW, over four times more than the amount of electricity the converter could pass through. This left large parts of the country in the dark while dealing with the worst nuclear disaster since Chernobyl.

The catastrophe did at least provoke change. In 2021, Toshiba Energy Systems and Solutions announced a new high-voltage direct current, or HVDC, converter, connecting the east and west grids. The power on the sending side is converted from AC to DC before being transmitted and returned to AC on the receiving side. Since direct current only travels in one direction, it does not bounce back and forth like AC, and so does not have a frequency – converting to DC allows the two sides with different frequencies to share electricity. The new converter increased the interconnection capacity to a total of 2,100MW.[9]

This isn't the only change that's been made. In the wake of the Fukushima accident, Japanese billionaire and technology entrepreneur Masayoshi Son founded the Renewable Energy Institute. Son is strongly convinced that renewable energy is the future, and has put forward a vision for an electrically connected continent, dubbed the Asia Super Grid. These electricity superhighways would connect Japan with surrounding countries, including China, Russia, Mongolia and South Korea, allowing the sharing and exchange of renewable energy between them.

What is surprising is that these electricity superhighways would rely on DC, rather than AC. While Tesla and Westinghouse's AC systems won out over Edison's DC cities in the 1800s, DC is now coming into play and paving a possible new future for electricity transmission. In Edison's time, it was not possible to step the voltage of DC systems up and down, so it travelled at low voltage and fizzled out over longer distances. But engineers continued to develop DC in the background. The advancements of power electronics and semiconductor devices in the 1970s were pivotal for DC, and it is starting to play a role in electricity transmission. As demonstrated in Japan, it is now possible to convert between AC and DC, and significantly increase the voltage of DC, something Edison could not achieve in his day.

Some tricks can be used to reduce the power that gets lost as heat when electricity travels across long-distance wires: adjusting the voltage and current. It all comes down to the relationships between voltage (the push on the charge) and current (the rate of flow of charge). At a very high level, doubling the voltage makes the losses four times smaller, so at a higher voltage, more of the electricity makes it to the other end.

While the relationship between voltage and current remains the same for AC and DC, there are differences between the two types of current that make high-voltage DC transmission preferable to AC. For example, a thinner cable is needed for the same distance and there are lower losses for DC transmission. The very first commercial example of this kind of DC long-distance transmission was actually in 1954, when a link connected Västervik, on the east coast of Sweden, and Ygne, on the island of Gotland in the Baltic Sea. Before this, residents of Gotland relied on a fossil-fuelled power station on the island to meet their energy needs, paying double the price paid by those on the Swedish mainland. It was expensive to run industries on Gotland, so people found themselves unemployed, and many left. Looking for a solution to this, the Swedish parliament decided to finance a pioneering project to reduce electricity costs and make life easier for the islanders. The original link transferred up to 20MW over a 100km cable, traversing land and sea, at a voltage of 100 kilovolts (kV). In the 1970s, enabled by developments in power electronics, the link was upgraded to deliver 30MW at 150kV.[10]

Any direct current transmission between 100kV to 800kV is generally classed as HVDC. But in a continent as colossal as Asia, there's call for an even bigger scale, known as ultra-high-voltage direct current, or UHVDC, with transmission lines operating at an even higher voltage – over 800kV. These lines are already used in China in particular, as the vast majority of China's citizens live in the east, in large cities like Shanghai and Beijing, while the country's power stations tend to be more than a thousand kilometres away, over on the other side of the country. The UHVDC long-distance lines carry the electricity over these vast distances into the urban centres.[11] For example, the 3,324-kilometre, 1,100kV Zhundong–Wannan line connects the deserts and mountains of Xinjiang province in the west to the Shanghai area on the east coast, with the capacity to move 12,000MW of electricity.

Converting power generated in AC to DC ready for transmission, and back again to AC at the other side of the cable, is expensive because of the equipment and operations needed for all these conversions. But when you are dealing with huge distances like this, as a continent like Asia has to, it does start to make sense – once you are sending anything over the distance of about 800 kilometres above ground, or fifty kilometres for submarine cables, converting to DC transmission starts to become cost efficient.[12] Super-scale transmission has the power to change things far beyond China, and could completely transform the global energy system, making it almost unrecognisable from how it is today. This is because it could alter the face of renewables; transmitting electricity at these high voltages could make large-scale renewable energy installations, like huge solar and wind farms in Mongolia's Gobi Desert, a possibility. Mongolia has the potential to generate 2.6 terawatts of electricity,[13] more than double the installed power generation capacity of the US, meaning that in the future, Mongolia, and other countries with similar potential, could become energy superpowers.[14]

But to realise this potential, Asia's electricity systems must be interconnected to move the electricity from where it is produced to where it is needed, to countries like Japan. This presents both a huge logistical challenge and an exciting opportunity for countries to collaborate. As well as completely shifting global power dynamics, increasing interconnectedness of electricity transmission systems is generally positive, as it can make the system more reliable. It's also good for increasing renewables in the system. More end users, and more diversity of when and how they use electricity, means intermittent renewable electricity is more likely to find a home.

These giant, interconnected systems were originally conceived to move electricity from huge, fossil-fuelled power stations to consumers – along an entirely one-way street. But as with any big problem, having multiple solutions is essential for the future of sustainable energy – and alongside the shift towards more interconnectedness, there is another, very contrasting shift taking place. In parallel to developments on an epic scale, there is a new era opening up for moving electricity small distances: microgrids.

*

Briceburg, Mariposa County, California, is home to a tiny community. It sits on a single winding road running alongside the Merced River, up to the El Portal entrance into the west of Yosemite National Park. Approaching Briceburg, the first thing you see is a visitor centre, a popular stop for tourists making their way to Yosemite. There are two houses – one is a holiday home, the other is Dave and Tracy Greenwood's home. Dave is a Bureau of Land Management River Ranger; he takes care of the nearby campground. Since moving to the area in 2000, Tracy has single-handedly managed the visitor centre, offering advice about rock climbing, rafting the river, hiking and backpacking, as well as occasionally helping campers during emergencies.[15]

The Greenwoods have seen some major natural disasters and accidents in the area over the years, including mudslides, rock falls and drownings, and in October 2019, the Briceburg Fire blazed its way through the area, burning over 5,563 acres of thick, dry vegetation.[16] The tiny communities were evacuated and the winding road was closed until the fire was brought under control more than two weeks after it started. Thankfully, no one was hurt, but the fire set a new course towards microgrids for this rural community and others similar to it.

A microgrid is a small version of the grid: a self-sufficient energy system serving a small energy-using community, like an industrial site, a university campus, a hospital or a neighbourhood. Microgrids come in lots of shapes and sizes, but generally tend to have a few sources of energy generation (for example, solar panels and diesel generators), a battery for energy storage, cables to distribute the power, and a controls system that acts as a brain to automatically monitor and control the supply and demand of electricity.

Rather than producing large quantities of electricity in a few power stations and pushing this out to large numbers of users through long transmission and distribution cables, microgrids take a more local, small-scale approach. This is more efficient, since shorter cables mean less loss of energy along the way. But one of the benefits of the large, interconnected grid is its reliability. If one power station goes down, another can usually step in to fill the gap, and the end users do not notice any difference. It is possible to have the best of both worlds, with a microgrid connected to the main grid, kicking in if there is an outage on the main grid and operating independently in what is known as 'island mode'.[17]

The Briceburg Fire, one of thousands of wildfires in California in 2019, destroyed the Pacific Gas and Electric Company (PG&E) cable to Briceburg, cutting off the community from their electricity supply. The Greenwoods and their neighbours were given temporary diesel generators, while

PG&E thought long and hard about what to do. Putting the long cable back in came with high maintenance costs, as it ran through about two kilometres of forest. Trees growing close to the electricity lines could catch a spark and set off a wildfire, which has happened in the past with devastating effects.

'I think everybody is tired of not knowing when we're going to have power,' Michele Nesbit told me from her home in Northern California.[18] Michele is co-founder and chief operating officer of BoxPower, a company that provides renewable microgrids. 'Everybody is starting to get a taste of that in California,' she said, referring to the power cuts in the state. When we spoke in July 2021, it was near the beginning of fire season, and there had been a few big ones already, according to Michele. 'For us, fire season just keeps on getting extended. PG&E don't want to add to that, so they're taking action to make sure that their lines are safe and down in the event of winds coming in.'

As part of PG&E's wildfire mitigation efforts, communities served by cables running through 'High Fire-Threat Districts' like Briceburg can expect to have their power turned off for days at a time, in what are known as Public Safety Power Shut-offs, or PSPS. Turning off the power line removes the risk of sparks from any cables running too close to trees, setting off a fire. With nearly 50,000 kilometres of electricity transmission and distribution lines running through high-risk areas, affecting nearly half a million households, mostly in rural areas, this is a huge problem for the utility company.[19,20] Instead of connecting Briceburg back to the main grid system, PG&E decided to give the community a microgrid, supplied by BoxPower. It is made up of solar panels, a battery and propane-powered generators as back-up, and it unleashes Briceburg from the grid and the power shut-offs.

'PG&E saw it as necessary action to put a microgrid there, not only for cost but also for resiliency, because we're in this time and place where forest fires are just going to continue getting worse,' Michele said. Removing the need for long cables reduces the probability of being cut off from electricity to prevent the risk of starting a fire, and also means a community is less likely to be cut off if there is a fire. This is something we are all seeing across the globe: more and more forest fires are devastating landscapes and communities, caused partly by climate change and the associated extreme weather conditions. The way we have used energy since the Industrial Revolution has made climate change worse, seemingly leading to more forest fires that destroy energy infrastructure – in a tangled way, it feels as if nature is fighting back.

The Briceburg system only took about three months to install, and Michele was there for all of it; from the first breaking of ground, up until the end when the system was turned on and the local press stopped by to report on the story. From Michele's home, Briceburg is four hours or so by car, so she stayed in a nearby Airbnb for part of the time, and at a nearby campsite for the rest. Luckily, Michele is a seasoned camper, which makes sense given she grew up down a dirt road in the middle of the woods in Grass Valley, California. No other houses could be seen from theirs, she tells me, and the nearest neighbours were her grandparents. 'When the weather would go bad, it was much more likely for us to lose power. It was almost a given, because the distribution lines were so long, [and] trees [were] falling over, taking the lines out,' she recalled.

Michele's family was used to dealing with these outages, making sure they had water on hand, as power outages meant no electricity to power the pump bringing water up from the well. She recalled having to put all the groceries outside in the snow to keep them cool one winter, during a multi-week power outage. 'I guess it's kind of like camping indoors,' Michele said. The US and Iraq could not be any more different on the surface – but despite the disparity in backgrounds and living under very dissimilar circumstances, I could easily relate to Michele's early experiences with energy, because they mirrored mine. The first-hand experiences and the pains of living with regular electricity outages motivated Michele to become an engineer, and she went on to study mechanical engineering and materials science, with a focus on renewables and sustainability. Looking back, perhaps my own childhood experiences with energy subconsciously led me to a career in this field too.

Michele and the team have installed solar and storage microgrids for many remote and rural communities, including Puerto Rico, after Hurricane Maria knocked out most of the power lines in 2017. This was the worst electrical blackout in US history, affecting the community for nearly a year.[21] Similar systems have also been installed for small communities in Alaska, who rely on diesel being flown in to power generators, paying a huge premium for it. One benefit of working with rural communities is getting to know the people there. In Briceburg, Tracy and Dave Greenwood regularly stopped by to chat with Michele and her team and check on them, helping out by taking in deliveries of parts ordered last-minute for the work. After living off diesel generators for a couple of years, the Greenwoods were thrilled to have this new microgrid.

The microgrid is oversized, to make sure the community has enough power. A remote monitoring and control system, via satellite, means PG&E and BoxPower can keep an eye on the system, respond to any warnings or alarms, and take preventative action to keep the power on. Briceburg could be one of hundreds of rural Californian areas to become untethered from the main grid and be powered by a microgrid, a huge shift from today's system.

'This opens up questions of the customer's concerns about reliability and the things they are losing by taking those power lines out. There's a lot that has to go into winning the customer over and changing the way that it's perceived,' Michele told me. Standardising these microgrids to make it easier to install and service them is key to successful deployment to rural communities living at the ends of long cables.

For the engineers and technicians used to looking after power lines, shifting to maintaining microgrid systems is a completely different ball game. Instead of sending out helicopters and people to walk the lines and inspect, they will need to maintain many smaller electrical systems. Checking filters, vents and fluid levels, washing solar panels and checking that no animals have nested in awkward places are just a few of the activities needed to maintain rural microgrids.

Microgrids may not completely replace the large grid systems that exist today, but they already play a role in certain settings, like Briceburg, Puerto Rico and remote Alaskan villages, and they can be low carbon. BoxPower aims for an 80% renewable fraction, so that at least 80% of the power is provided by solar over the course of a year, enabled by the battery storage element in the microgrid. Any shortfall in power demand is met by the back-up fossil-fuelled generators (propane, in the case of the Briceburg microgrid).

Looking ahead to the next hundred years and beyond, one thing is certain: we will continue to use electricity, and therefore will need to move it from where it is generated to where it is consumed. That may be through ultra-high-voltage directcurrent mega-transmission lines across continents, or it may only be a cable of a few metres, connecting one or two buildings to a microgrid. From huge to tiny, both of these options can get us off fossil fuels and onto renewables. Huge interconnected systems provide a diversity of end users and, combined with smart grid management, enable the use of renewable energy generated at times when the sun shines or the wind blows.

Tiny microgrids, especially those with batteries, can store electricity from renewables for later use. In the future, all these options could play a role, with different systems suiting the needs of different geographies, terrains and populations. But to really transform the way we use electricity, and more generally, whole energy systems, we need both a greater variety and better ways of storing energy.

Chapter 8

Energy Storage

THROUGH THE TREES, Cara and Trevor's wooden summer house revealed itself, looking just as scenic as I remembered from my last visit. Crowned with a gabled roof, it is perched on top of a hill on the east coast of Canada, with views over St George's Bay and the Northumberland Strait, spilling into the Atlantic Ocean. The setting sun's rays over the horizon scattered oranges, pinks and purples over the clouds and water. It's not hard to see why they built their home here; set in total tranquillity, surrounded by the scent of fresh wood and magnificent scenery.

Cara is my husband's cousin, so I have visited their home in the wilds of Nova Scotia – Latin for 'new Scotland' – a few times. The Scottish heritage of the region is reflected by page after page of MacDonalds, MacLeods and MacNeils in the local phone book, annual Highland games, and an unusually high presence of bagpipes. Cara and her husband Trevor spend their summers there, and the rest of the time in Calgary, in western Canada, where they work. 'We decided to build the house so we could spend more time in Cape George and spend more time with Mom and Dad and family in general,' Cara told me.[1] The Cape is a tiny fishing community, each plot of land inhabited by extended families: aunts and uncles, cousins, and friends who have been around for so long they are pretty much family. Everyone knows everyone, but newcomers like myself are welcomed in with very little suspicion.

Being at the top of the hill comes with amazing views, but it also comes with an energy problem. While most homes in the area are connected to the local electricity grid, it does not stretch far enough to reach Cara and Trevor's house. The cost of connecting their home to the grid would have been around $100,000 in Canadian dollars, or £60,000, prompting them to look into alternative options when they built the house in 2014. They decided to stay off-grid and looked into possibilities for generating their own energy to become self-sustaining. After exploring a number of options, they ended up joining a local solar panel co-operative, which solved most of the problem. But while solar panels convert energy from the sun into electricity throughout the day, they of course become useless after the sun goes down. So, Cara and Trevor found a clever workaround.

After the glorious sunset was over, I snooped around a neat white wooden shed next to the house. The roof of the shed is tilted at about forty-five degrees and is topped with twelve sunbathing solar panels. Inside the shed is Cara and Trevor's ace of spades: sixteen lead-acid batteries. These rechargeable batteries store any excess energy generated by the solar panels during the day, for Cara and Trevor to use in the evenings, making the most of the natural resources available to them.

Tucked away behind their house, there is another form of energy stored: two fat, white, cylindrical tanks store propane, a petroleum product, which provides their fuel for heating water and the stove. A solar and battery system large enough to provide all the energy needs for the home was, in this case, prohibitively expensive, leading to this compromise. For their life off-grid in this fairly remote part of the world, being able to store energy is essential – without it, Cara and Trevor would be left in darkness the moment the sunlight fades.

Cara and Trevor's lead-acid batteries use the electricity to power a chemical reaction that can later be reversed to release electricity when it is needed – reversible chemistry being the key to storing energy in a battery. A lead-acid battery is the type found in most petrol and diesel cars (although Cara and Trevor's version is the type more commonly found on boats), made up of lead plates submerged in a liquid acid, hence the name.[2] Two plates, one made of pure lead, and the other of lead oxide, react with the acid, releasing electrons. A conductor, such as a copper cable, connects the two plates externally to the acid-filled battery, allowing electrons to flow between them – a flow of electricity. If that copper cable passes through devices, like Cara and Trevor's refrigerator, the flow of electrons will power it up. Eventually, the chemicals are used up and the reaction stops, meaning no more electrons flow. But a supply of electricity from the solar panels to the battery will reverse the reaction and recharge the battery, allowing the chemical reaction to take place over and over again.

Well-established lead-acid battery technology tends to come at a lower cost than other more modern and popular options, such as lithium-ion batteries, the type that now power our lives as they are essential components of electric cars, mobile phones and laptops. That said, the cost of lithium-ion batteries is very quickly dropping as the technology develops, more resources are discovered and giant factories to make them pop up across the world. These batteries are more versatile than their older lead-acid relatives, working at both a tiny scale when storing energy for our portable devices, and at a large scale, storing excess energy from renewables. My fifteen-year-old self's mind would be blown if she could see the plethora of battery-powered devices that make my life more efficient and convenient – a mobile phone constantly connected to the internet, wireless headphones, electronic reading devices containing hundreds of books at a time and cordless vacuum cleaners.

Lithium-ion batteries work using the same principle as their lead-acid relatives – a reversible chemical reaction releases electrons – but they use different materials to kick off the reaction: lithium-metal oxides, graphite and solvents. In 1991, two Japanese companies, Sony and Asahi Kasei, worked together to release the first commercial lithium-ion battery. This is surprisingly recent, given how ubiquitous they are today.[3,4] But, as with anything ingenious, there are still kinks to be worked out – for example, how to reduce the weight of these heavy batteries in electric vehicles, so that more of the energy they provide is used to move people and goods instead of the heavy battery itself. There are some much more serious issues that stretch beyond functionality, namely humanitarian and environmental factors. As with most technologies, we have to think of the totality of the solution and the available resources.

The demand for lead-acid batteries is still rising: they are used for motorised cars, for storing electricity generated from renewables, and for back-up power supplies, which is resulting in increased demand for lead. The good news is that these batteries are heavily recycled, so a significant proportion of the demand can be met with recycled lead. But lead is a very toxic heavy metal, and any exposure to it affects almost the entire body, including the nervous system – lead poisoning can disrupt brain function, especially in young children. With the right safety systems and protective equipment in lead-acid battery recycling facilities, exposure can be avoided, but this is not always the case.[5] 'Backyard' or 'cottage' recycling happens across many countries like Senegal and Vietnam, with little pollution control. The work may be done at home by families, with children helping to dismantle batteries and wash the components. Parents may lack knowledge of the toxicity of lead, or they may simply not have any other employment options. The issue is not limited to small operations – Exide Technologies, a producer, distributor and recycler of lead-acid batteries, was forced to close down its large California recycling facility in 2015 after contaminating around 10,000 homes with lead, some as far as three kilometres away from the facility.[6]

The lithium-ion battery story is not complete, because they have not been around long enough. Their full impact is yet to unfold as we navigate through the full lifespan of this type of battery. By 2030, we could see 300 million electric cars on the road, thirty times more than in 2020,[7] which means we have to find more of the ingredients needed to make batteries – lithium, nickel, manganese and cobalt. And we will need even more if batteries are built to store electricity from renewables to support the grid.

Lithium is not particularly scarce, but getting it out of the ground comes with environmental problems. South America's

'Lithium Triangle', spanning Argentina, Bolivia and Chile, is thought to hold a large proportion of the world's supply of lithium under its salt flats. The biggest obstacle in its extraction is water; around 2,000 litres of water are needed to extract every kilogram of lithium, in one of the driest regions in the world.[8, 9] In Australia and China, where much of the rest of the world's lithium supplies come from at the moment, it is mined as a mineral called spodumene, and processed at temperatures of over 1,000 degrees Celsius, demanding huge amounts of energy.[10]

A few organisations are looking at more sustainable ways to mine lithium. One alternative is to extract geothermal brine, a hot, concentrated salty solution that has been circulating through hot rocks and picking up elements, including lithium. Nature has done some of the work already, using geothermal energy to sap the lithium out of the solid rock. The rest of the heat needed to power the processing of the brine to get the lithium out could come from the geothermal resource. If it works at a large enough scale, this method could prove particularly useful in the Rhine Valley in Germany, a global hub for car manufacturers. Rather than shipping in lithium from afar, adding to the carbon footprint, lithium extracted here could be passed straight to the car makers.[11]

Cobalt, one of the other metals used in Lithium-ion batteries, is even more problematic. As well as the energy needed to extract it, the majority of its reserves are in the Democratic Republic of the Congo, which has been widely condemned for human rights abuses associated with the mining of this material.[12] As our demand for lithium-ion batteries increases, we increase the demand for cobalt, and in doing so contribute to a system that has been linked to abuse and child labour – unless we find an alternative. Researchers are investigating the possibility of designing out the need for cobalt in batteries altogether. Experiments are ongoing with alternatives like manganese and iron-based materials, with low or even zero-cobalt iron alternatives already available, making use of materials that are more abundant and easier to get to. If the development of alternatives doesn't quite replace existing practices, there is a possibility that countries with large amounts of these metals – lithium, manganese, cobalt, nickel and others needed to manufacture equipment for the energy transition – could become superpowers, overtaking today's fossil giants.

Once you've considered the difficulties in producing batteries, you also need to consider the end-of-life options for them. Batteries don't last forever, and after years of charging and discharging, they can lose their effectiveness. Anyone with a mobile phone will recognise this frustrating gradual decline in battery life. Recycling lithium-ion batteries is difficult and has been likened to trying to get flour out of bread. But it is not insurmountable, and thinking about the end of the life of a battery right at the beginning can and should influence the design choices made throughout the process in order to make recycling possible. Ultimately, legislation will have to drive this, and governments are starting to take positive action.[13] In December 2020, the European Commission proposed new plans to make sure batteries are produced with the 'lowest possible environmental impact, using materials obtained in full respect of human rights as well as social and ecological standards. Batteries have to be long-lasting and safe, and at the end of their life, they should be repurposed, remanufactured or recycled.'[14] It seems, on paper at least, that we are heading in the right direction.

For many of us, energy storage is a luxury for devices and toys. But for others, such as those who are off the main electricity grid, like Cara and Trevor, it is an unavoidable necessity. Whenever I visit Cara and Trevor's home, it's clear to see how pairing renewables with batteries enables them to power up their daily needs, from the lights to the refrigerator, and most importantly the pump for the well, which supplies them with fresh water. However, most of us still rely on fossil fuels, which don't have any of the same storage issues as renewables. The fact that it is easy to store fossil fuels, allowing a constant and reliable source of energy, is perhaps one of the reasons for our reluctance to wean ourselves off them. A very simple version of energy storage would be sheds full of an arguably renewable source of energy – chopped wood, an essential for many homes in Nova Scotia that rely on wood burners to get them through the winter. At the other end of the scale, colossal oil and LNG ships are also a form of storage, storing the fuel while in transit. Some of the ships actually end up moored in one location towards the ends of their lives, becoming permanent storage tanks.

A more complex example comes in the form of underground oil grottos, which were created by the US government following the oil shocks of the 1970s to store an emergency supply of oil in the event of any future shocks. The grottos, located on the Gulf Coast, are salt caverns that were hollowed out by drilling a well into an underground salt formation, and injecting fresh water to dissolve the salt and carve out a cavern. Precise dimensions were created by carefully controlling the water injection and extracting the salty water. The resulting salt caves are impermeable and non-porous; in other words, each one is a waterproof and sealed storage tank for oil, with self-healing walls that seal up any small cracks that develop. Oil can be moved in and out of these caverns because it floats on, and remains in a separate layer to, water. To get oil out, fresh water is pumped into the bottom of the cavern, displacing the oil and pushing it out to the surface. The opposite happens when oil is pumped in for storage, displacing the water.

Four of these underground grotto oil storage facilities exist in Texas and Louisiana, ranging in size from about a million to six million cubic metres. A typical cavern is about as tall as two Empire State Buildings stacked on top of each other. The four storage facilities are home to sixty of these gigantic underground caverns, strategically located with access to pipeline distribution systems to oil refineries, interstate crude oil pipelines, and marine terminals for shipping crude oil.[15]

Oil and gas are not only stored to protect against energy crises, but also to deal with changing demands across the seasons; this is called 'inter-seasonal' storage. In Europe, to generate electricity and heat, two or three times the amount of natural gas is needed in the winter compared to the summer, especially in the colder north. But natural gas is produced all year round, so storage is needed to hold onto it through the summer until the demand goes up in the winter. Excess gas can be stored in salt caverns, like the ones that hold the US's emergency oil reserves. An alternative is to store gas back in the rock formations it once came from – in the empty space left behind from gas extraction.[16]

The act of storing energy relies on our ability to hold or convert it into a molecule that can be stored and used later to release energy. In a future fossil fuel-free world, based only on renewable energy, hydrogen could be used as one of those storage molecules. The generation of renewable electricity, like wind and solar, does not always coincide with when the electricity is needed. If it is not immediately used, it will go to waste. But it can be used to power the process of electrolysis, splitting water into its constituent parts of one oxygen molecule to every two hydrogen molecules. Hydrogen fuel can be consumed immediately, or it can be stored underground, like natural gas. When needed, the hydrogen can be recombined with oxygen in the air, releasing water and energy.

Hydrogen has, in fact, already been stored in salt caverns in some places, including the US and the UK, but the stores have not yet been fully tested to see how they cope with quick and variable injection and extraction of hydrogen. This question is one of many being addressed by research projects; others are investigating if hydrogen is lost during injection and extraction, and if so how much, and whether any underground microbes will decide to feast on the hydrogen and release their wastes into the mix.[17] Until other options are mature enough, batteries – despite their problems – do offer the possibility of a revolution in energy storage that could be used to clean up electricity generation and coax us off fossil fuels in the long run.

What can we, as consumers, do to make sure we continue on this trajectory to embrace newer advances in energy storage without doing additional harm? Some of the issues can be overcome by consumers questioning what materials are in our devices, and paying attention to how they have been made. We should demand the creation of safe, secure, environmentally aware mining operations, where workers are treated fairly – and we must be prepared to pay higher prices for this. Alongside this, pressure is needed from consumers to force manufacturers to collect and publish data on these issues, so that we can make informed choices. Crucially, batteries need to be designed with the end of their lives in mind, so that they can be reused and recycled to make the most of the resources we have, minimising the need to mine fresh materials. Some people have suggested mining asteroids and the deep oceans to find fresh resources, but it's only by focusing efforts on better management of available resources that we can avoid damaging even more of our planet and universe.

*

Electricity is tricky. Unlike other commodities, like petrol, water, coal or wheat, it cannot be stored in a tank or a silo, waiting until it is needed. For this reason, it has traditionally been used as it is produced, with the likes of the National Grid commanding power stations to be turned on or off to balance supply with demand. While some methods for storage of electricity, like lead-acid and lithium-ion batteries, do exist in smaller pockets of the energy system, in the grand scheme of things, there are not many. Without a method for large-scale storage of electricity, enough generating capacity is needed to meet the highest possible demand, which means planning for the worst-case scenario. For the majority of the time, though, this leaves lots of spare capacity: power stations that only get used at the extremes – the coldest or hottest days of the year, or when something unexpected happens. Not only that, but on windy and sunny days, the electricity generated by renewable wind and solar farms can go to waste if there is not enough demand during those hours.

Storing electricity on a larger scale allows us to maximise the use of renewable generation. It also avoids the need for huge power stations, designed and built to meet the peak demand, as stores of electricity can be used to top up supply when the demand is highest. Since energy can be converted from one form to another, electrical energy can be transformed into other energy types. Moving up the storage scale, one fairly efficient way to do this is to use the electricity to pump water uphill, store it in a reservoir where, because of gravity, it has potential energy, and then release it back down the hill through turbines, spinning them as and when the energy is needed. In this instance, the electrical energy has been converted to potential energy, which is then converted to kinetic energy to drive the turbine and generate electrical energy again. The process is very much like the Three Gorges Dam power station in China, except that water is forced uphill rather than relying on the natural flow of a river. Unlike lead-acid and lithium-ion, this type of storage relies on mechanics rather than chemistry, but the underlying principle is the same: you convert energy into different forms that suit your needs.

This is precisely what happens at Cruachan Power Station, in Argyll and Bute, on the west side of Scotland. The power station takes advantage of the natural landscape, pumping water from Loch Awe, a long and thin freshwater loch (the

Scottish word for a lake) up to an artificial reservoir close to the peak of Ben Cruachan, a one-kilometre-high mountain towering above the loch. Compared to the little white shed of lead-acid batteries in Nova Scotia, this is a mammoth energy storage operation in old Scotland.

Scottish legend has it that Cailleach Bheur, a one-eyed giant with white hair, dark blue skin, and rust-coloured teeth, was responsible for guarding a well at the top of Ben Cruachan.[18] Every night, she covered the well with a slab of stone, lifting it away the next morning. One unfortunate evening, she fell asleep and neglected her duties. The well overflowed and water ran down the mountain, creating a new outlet to the sea through the Pass of Brander. This is how the River and Loch Awe formed. The old woman was turned to stone as punishment and sits above the Pass of Brander to this day.

When I found myself driving along the road that runs between the loch and the mountain, the blue one-eyed giant crossed my mind. The area was peaceful, still and empty. The only clue that there might be something more afoot was an electricity cable up in the distance, strung out between a number of pylons among the trees, running up the side of the mountain. I craned my neck up, curious to find the top of the mountain. A green sign at the entrance read 'Cruachan', but there were only normal office buildings to be seen, and no indication that this was a power station. If you were to slice into the mountain, however, a different story would appear; hidden deep inside the rock is a cavernous machine hall, giving it the nickname the 'Hollow Mountain'.

The idea for Cruachan Power Station originated in the 1930s, the brainchild of Scottish engineer Sir Edward

McColl, who was a pioneer of hydro-electricity in Scotland. After getting permission from the government to build it, and once the design was complete, construction finally started in 1959. Thousands of men worked on the project over a six-year period, before the power station was opened by Queen Elizabeth II in 1965.[19] A one-kilometre-long tunnel, leading to the machine hall in the centre of the mountain, had to be hand-drilled and blasted through solid granite. The post-war project attracted migrant labourers from Ireland, Poland and other areas. The money was good, described as 'football wages' by Sarah Cameron, the Cruachan Visitor Centre Manager, but the job was dangerous and working conditions were difficult.[20] The 'Tunnel Tigers', as they are affectionately called, worked twelve-hour shifts, and some even worked 'ghoster' shifts of up to thirty-six hours in a row. They drilled holes into the granite with handheld drills, packed the holes with explosives, waited for the blast, and removed the rubble continuously for six years to excavate the access tunnel, turbine hall and the penstock – the shaft that connects the upper reservoir to the turbines. While the Tunnel Tigers dealt with a hot, dusty environment inside the rock, at the top of the mountain, a separate construction project was taking place to build a buttress dam and create the upper reservoir. The workers at the top of the mountain had to deal with harsh winter weather, and health and safety practices were not what they are today – fifteen men sadly lost their lives during the project.

Travelling down the access tunnel is a unique experience, to say the least. The entrance sits behind a metal gate, accessible only to those working at the power station. In this case, it was Ian Kinnaird, Scottish Assets and Generation Engineering Director at Drax, a power generation business, who took me and my husband as visitors in his car. The slow drive through the dimly lit, damp and misty tunnel felt like the journey into a Bond villain's lair. The curved surface of the tunnel was rough, dimpled with the original hand-drilled marks left in the granite. At the point where we couldn't go any further by car, we continued down the tunnel on foot, machinery gently humming in the distance. Now deep into the mountain, we passed some pot plants, which looked very out of place – noticing my confusion, Ian and Sarah explained that these had been part of an experiment to see what would grow down here, under an ultraviolet light (it turned out anything with green leaves and white flowers did well, but coloured flowers did not). As we continued along the path, the humming got louder, increasing in tandem with my anticipation for what lay at the end of the tunnel.

The machine hall ahead took my breath away. My eyes were met with a gigantic room – the size of a football pitch with a curved ceiling higher than a ten-storey building. I took in the detail; a beautiful two-tone cream-tiled floor stretched across to four large yellow barrels – these must be the turbines, I thought to myself. To my left, a wooden mural filled a large space across the wall, put together by artist Elizabeth Falconer to tell the story of the power station. In the bottom left corner was Cailleach Bheur, the old woman from the legend. A Celtic cross dominated the middle of the picture, and fifteen faces signified the men who lost their lives during construction. Four tall characters represented the men who managed to get the project off the ground, and Queen Elizabeth II could also be seen commanding the power station into action. The last section, on the right-hand side of the mural, showed the dam, the tunnel, the machine hall and Loch Awe. It was certainly an unexpected artwork to find in the centre of a mountain.

Below us and to the left, I spotted some windows into the control room. 'Someone is always underground in that hole,' Ian told me. 'Operators have staffed this room 24/7, 365 days a year, since the opening of the power station, working twelve-hour shifts in pairs.' Down on ground level, standing near the control room windows, the four turbines towered above us, the four beating hearts of this entire operation. Designed to be reversible, they can pump water uphill, and also serve as an electricity generator, spinning the opposite way when the water rushes back down through them. This was a world first when it was originally built – a one-of-a-kind pioneer. Other similar pumped hydropower stations had separate pumps and electricity generators, so it takes them around an hour to turn bits of equipment on or off to switch from pumping water uphill to generating electricity, but Cruachan can do it in less than ten minutes because of the reversible design.

When all four generators are going at full throttle, they can produce a total of 440MW, enough for the electricity needs of more than 90,000 homes for a year.[21] Originally built to complement nuclear power stations, Cruachan plays a slightly different role today. Like renewable electricity, nuclear power is also relatively inflexible, unable to easily ramp up or down its electricity output in response to demand. Cruachan came in to provide flexibility – by absorbing the excess energy generated by the nuclear power stations overnight, when demand was low, and releasing it during the day.

Nuclear power stations are predictably inflexible, whereas renewable electricity generation varies with the whims of the weather. Nowadays, Cruachan's flexibility has been applied to renewable electricity. It can start to generate electricity in under thirty seconds, so can be called on by the National Grid to fill in any last-minute gaps; perhaps clouds obstructing the sun or sudden low wind speeds. On the other hand, if there is over-generation of renewable energy – say, on a particularly sunny day when fewer people are turning on lights – Cruachan can start to use the electricity to pump water up the mountain less than seven minutes after being asked to do this, storing it up in the reservoir until electricity is needed again. Because it is reversible, the operators can change Cruachan from generating electricity to pumping water, starting and stopping multiple times a day and thousands of times a year.

Travelling back out of the hollow mountain, through the hand-made tunnel, we next made our way to the dam up a steeply winding road. En route, we passed by a shepherd's cottage with a tiny boat sitting in the front garden; the word 'TITANIC' was printed in white across it. Sheep grazed, and a sign warned us: 'Baby lambs, slow down'. The pylons and cables I had spotted from the road down below were now a lot closer, and yet still, they were the only giveaway that something unusual was happening here. Around 396 metres above Loch Awe, the dam came into view – a view I was later surprised to spot on television, while watching the *Star Wars* series *Andor*, for which it was transformed into a scene in outer space. For any 'Munro-baggers' out there, three can be captured by hiking Ben Cruachan. For those unfamiliar, a 'Munro' is a term for a mountain in Scotland above a certain height (914m), named after Sir Hugh T. Munro, who surveyed and catalogued them in 1891 (and 'bagging' is what the Scottish call climbing them).[22] I imagine even the most seasoned Munro-bagger would be shocked to turn a corner in this quiet rural idyll and come face to face with a huge dam, which stretches for 316 metres, holding back ten million cubic metres of water when the reservoir is full. That volume equates to about seven gigawatt hours of usable, stored energy, which can be dispatched over a continuous sixteen-hour period if needed. The reservoir also pulls in about eighty-five gigawatt hours per year of free energy, which is collected as rainfall through tunnels into adjacent valleys, leading the rainwater to the reservoir, in addition to the water that is pumped uphill.

A series of hidden tunnels and pipes, like veins and arteries, connect Loch Awe and the reservoir to the generators, the beating heart of the operation. A pair of penstocks, 260-metre-long inclined tunnels (picture these as giant water slides), bring water down from the reservoir, each pipe splitting into two and connecting to one of the four turbines. In the past, someone would be sent down these penstocks to inspect them on a scaffold with wheels on it – an undoubtedly unnerving job – but nowadays it's done by a robotic remotely operated vehicle.

Another one-kilometre-long tunnel runs parallel to the access tunnel but is even deeper underground. Water is either drawn in through this tunnel or released back out to the loch after it has passed through the turbines and generated electricity. Left to its own devices, this would result in rumours of mythical creatures to rival the Loch Ness monster. Drawing water in or letting it out would naturally create a whirlpool, like the vortex seen as water drains down a sink, or a gurgling fountain in the loch. But this tunnel system has been designed with a surge tank that fills up with water and controls the flow in and out of the loch to avoid a swirling vortex. The final ventilation and cable tunnel, cored vertically down through the mountain to meet the machine hall, is home to the cables that bring the electricity up to the transmission cables atop the pylons by the reservoir.

Cruachan was designed to survive for forty years, but it has already outlived this, and will continue to operate for the foreseeable future. The team at Cruachan have even applied for permission to build a second pumped hydro plant alongside the existing one, which could be up and running by 2030, adding another cavern and more than doubling their electricity output.[23] Their success is part of a wider global renaissance of pumped hydro. As more and more countries lean into renewable energy, more storage will be needed to enable its generation. Australia's Kidston pumped hydro storage project, located on a former gold mine, got the go-ahead in 2021; it will be the first one of its kind built in the country for over forty years, and is set to become the country's third largest electricity storage device.[24, 25] In Israel, the federal government has plans to build 800MW of pumped hydro storage.[26] Some South American countries like Brazil and Colombia are also constructing new pumped hydropower stations several orders of magnitude larger than Cruachan to fit their needs.

Using the highs and lows of the land and the available water to store energy is a clever way to work with nature, but it depends on the topography of the place. Pumped hydro will only work in specific geographies, which is quite a limitation for this storage method. It also comes with risks; without strong environmental regulation, important habitats could be destroyed in the process of flooding an area to create the reservoir. Not only that, but as with the Three Gorges Dam, there are instances of entire populations being relocated as their villages are flooded for the purposes of a hydro project. There are always many sides to a story, and while this method of storing energy can help reduce harmful emissions, no one solution is a magical fix.

Large ships sailing the oceans, small trucks carrying fuel along windy roads, ultra-high-voltage power lines, tiny microgrids, or pipelines snaking through land and sea – these all play a part in the energy movement game today, and will evolve to be part of its future too. Likewise, each type of storage technology could play a different role in years to come. Small-scale batteries in electric vehicles could provide a storage service to the grid. Cara and Trevor's solar and battery system is a match made in heaven for off-grid homes, but a few years down the line, they may be able to switch out their propane tanks for hydrogen or another low-carbon option. On a larger scale, pumped hydro or hydrogen stored in salt caverns could step in to support the electricity grid daily, or to store energy over the summer ready for the winter.

We have to use all the technology tricks up our sleeves to switch the energy system from fossil fuels to renewables with the ultimate goal of getting it to the end users: you and me. With all this precious energy, we will go on to heat and cool our homes, we will manufacture an astounding assortment of objects, from cheap plastic toothbrushes to high-end electronic devices, and, of course, we will burn through energy to transport ourselves and our goods around the world.

The journey that energy will go on to reach end users is only one part of the story and plays only a secondary role in tackling the impact on our environment. The reason these gargantuan efforts to generate energy happen at all is in order to meet the energy demands of individuals and societies. So, the next big question is, once we have this energy within grasp, how do we use it?

PART 3

USE IT

Chapter 9

Heat and Cool

I SPENT MONTHS on the hunt, like some kind of engineering detective. In my bright yellow and red overalls and steel toe cap boots, perplexed bankers eyed me with interest, momentarily distracted before rushing to their next meeting. In the heart of London's financial district, I was on a mission: following a pipeline. Unlike my fascination with the shiny yellow gas pipes being installed around the city, this time the obsession was part of my actual job.

The pipeline was about six kilometres long, and I was following the entire thing from start to finish. My mission began at Citigen power station in Farringdon, where the pipeline left the building and dropped below street level, travelling underneath Smithfield Market, London's famous centuries-old meat market. As I wandered through at around midday, the market was deserted – it springs to life when most of us sleep, between two and six o'clock in the morning. Even though there were no visible signs, I knew roughly where the pipe was thanks to the engineering drawings I had examined that morning. I skirted the edges of the market, sketching a mental picture of where the pipe below me ran.

From here, I knew the pipe continued travelling deep underground before it eventually shot back up less than a kilometre away, running along the ceiling of a car park and into what was then the location of the Museum of London, by the iconic Barbican Estate.[1] It then dived back down beneath the ground into the premises of a law firm and snaked through other surrounding buildings. It entered and exited the mechanical rooms, normally hidden away in basements and filled with the equipment to control a building's heating, cooling, lighting and security systems. I got to know the caretakers of these buildings over time, as they helped me to access these rooms and shared their knowledge.

The pipeline I was following was part of a district heating network, a method of heating buildings common in Scandinavia, Germany, Russia and a few other countries, but unusual in others. In countries with bitterly cold winters, keeping buildings warm is a vital necessity for the people inside them, and a big user of energy. In the case of Citigen power station, a small gas-fired engine burns natural gas to generate electricity. Depending on the specifics of the power station, generating electricity in this way is only about 40–60% efficient, so around half of the energy contained in the fossil fuel is lost, escaping into the environment as heat. This is obviously hugely inefficient, so in some situations heat can actually be captured and put to good use. District heating systems can use the waste heat to warm up water, which travels through pipes like the ones I got to know so well, delivering heat into residential and commercial buildings dotted around an area. The pipes in this system were old and regularly sprang leaks, some more dramatically than others. Not only is this a waste of water, but it's also a huge waste of the energy consumed to generate the heat. Small, seemingly innocuous leaks can escalate to catastrophe with little warning, spurting boiling hot water and flooding basements. Finding and fixing leaks was a vital part of my job to keep the heat network functioning.

Rachael, a sharp-minded mechanical engineer and my boss at the time, introduced me to this particular leak. I had rolled myself over to her desk on my wheely office chair and looked over her shoulder at the computer screen. 'We know we're losing water from the hot water district heating pipe, but we don't know where it's leaking,' she said, pointing at the numbers on the screen in front of her. The dark screen showed a simplified line diagram of the pipes, with live information collected from sensors that measured the 'flow' and 'return' water volumes – numbers that should match. Hot water leaving our giant heater was supposed to travel around the six kilometres of pipe, giving away its heat via heat exchangers at each building. Heat exchangers are a way of passing heat between two fluids without mixing them together; the heat passes from a water pipe to the water system of the building while the two fluids remain separate. This is a bit like warming up someone else's cold hands with your own – you clasp their cold hands in yours, passing over the heat carried in your arteries and veins. As heat is passed from hot to cold, the water inside the district heating pipe cools down, and the cooled water returns back to the power station to be reheated and repeat the cycle.

But, as Rachael pointed out to me, about ten cubic metres of water a day was vanishing, which was equivalent to leaving a garden hose constantly half turned on. This isn't a huge amount of water to lose in one day, but no one could figure out where it was coming from, so it had now been leaking for six months. Compared to other problems, like burst pipes or engine troubles, which needed the engineers' immediate attention, this seemed like a minor issue. But the total amount of water lost was rapidly approaching 2,500 cubic metres, the amount needed to fill an Olympic-sized swimming pool, and as a new recruit, I had the time to focus on it while others dealt with more pressing problems.

Finding the tiny hole responsible for the leak was like looking up at the night sky for the pinprick of light of a specific star. I talked to the power station's engineers, technicians and operators, gathering intel and evidence. Some had worked on the system for many years, and were familiar with its habits and quirks, telling me to check on mechanical rooms they thought were behaving unusually. For several months, I followed many leads and tried different leak-detection techniques, to no avail. But just as I was about to give up, I went back through all of the information I had gathered and looked for connections – and suddenly there it was. All the clues pointed to one spot.

I walked over to the Museum of London and into their tiny mechanical room. As I stood there, I felt a warmth radiating from the wall on my right. I placed my hand on it, and it was warm to the touch. I also noticed a strange curvature; the entire wall had a worrying concave bulge. I confirmed my suspicions by arranging for some specialists to use leak-detection equipment to listen for the leak. The offending section of the pipe was finally revealed, removed and replaced.

Despite being a relatively small problem on the district heating system, given its age and complexity, I did feel a sense of satisfaction at not giving up and persevering for months to solve this puzzle. And it meant the problem was dealt with before becoming bigger, more expensive, and more dangerous – somewhat like the issue of climate change, which over the years has compounded into an enormous threat. As a new engineer, I learned the value of listening to everyone around me, and respecting their instinctive knowledge and wisdom. Without the input of the operators, I wouldn't have been able to put all the pieces together. My dedication to solving the problem did not go unnoticed; among my colleagues at the power stations, it became affectionately known as 'Yasmin's leak'.

*

We rely on nuclear power stations, wind farms, pipelines, cables and people like me to do their job properly so that we can heat or cool our homes, cook our food, and turn on lights. On gloomy winter evenings, the heating systems in our homes kick in and radiate warmth; aside from thinking about paying the energy bills, I would guess that the majority of people rarely think about how this even works in the first place. But turning heating on uses enormous amounts of energy, most of which result in greenhouse gas emissions.

In the European Union, four-fifths of the energy used in homes is for heating.[2] As warm-blooded creatures, a drop in internal temperature wreaks havoc on the chemical reactions that take place inside our bodies to keep us alive and well, like breaking down food or building up new cells. Over time, heating methods have evolved from burning wood in a simple open fire, to fireplaces and stoves, to elaborate central heating systems. Without these, many parts of the world, such as those close to the Arctic Circle, would be uninhabitable. How buildings are heated plays a big part in the overall energy system, and consequently its contributions to greenhouse gas emissions and climate change.

In cold countries, heating is responsible for the biggest portion of energy consumption in homes, and many have their own central heating systems that are made up of three key building blocks: an appliance that burns a fuel to generate heat; a medium (like air or water) that picks up that heat and moves it around in pipes or ducts to where it is needed; and another appliance that emits the heat into the space being heated.[3] In the UK house I grew up in, we had a boiler that burned natural gas to generate heat; the heat was picked up by water in a closed loop of pipes; and that hot water entered the radiators in each room, designed to let go of the heat in the right place to warm up the room. The water makes its way around the loop repeatedly, picking up heat from the boiler and letting go of it at the radiator. The cooler water then returns to the boiler to gather more heat. This system tends to be common in countries that benefited from a boom in natural gas production from the North Sea, like the Netherlands, where buildings are connected to a national network of pipes, supplying them with natural gas.

In other parts of Europe and the US, homes are connected to their own national networks of natural gas pipes, but the gas is burned in a furnace rather than a boiler. In these systems, the heat from burning gas warms up air, which is forced through ducts and blown out into the rooms being heated. It is also possible to use steam as the heat carrier – this involves water being heated up to its boiling point, turning it into a gas. It takes more energy to generate the steam, but it is more efficient at transferring the heat, so it works well in extreme cold or in industrial settings across the world.

District heating systems are not a new way of heating homes and buildings – in fact, one in Denver, Colorado, in the United States, has been going for over 140 years. Shadowed by the Rocky Mountains, on a crisp day in downtown Denver, the air temperature hovers just below freezing. While there are no signs of it at street level, several metres down, a network of pipes snakes around, carrying hot steam into the buildings. This heat supply has been keeping Denverites warm since 1880.[4] It is the oldest continuously operating system of its kind in the world. Before the existence of this district heating system, buildings were heated with individual wood-fired boilers, as was the case for Denver's historic Gumry Hotel. Every evening at the hotel, a young engineer had to clean out the day's ashes from the boiler, and fill it up with enough wood to keep it going for the night. Scraping out ashes in the hot and dusty environment was not a pleasant job, and playing with fire came with inherent dangers. One night in 1895, just after midnight, the hotel's boiler exploded with the force of a bomb, demolishing the building and killing twenty-two people. The *Aspen Weekly Times* reported that 'naught but the walls were left intact'.[5] The engineer on duty that night, Elmer Loescher, was blamed for the explosion and accused of being drunk on the job.[6] Fearing for his life, he fled for Southern California, but only made it as far as southern Colorado before getting arrested on charges of manslaughter and criminal carelessness. However, by the end of the trial, the jury on the case pointed out that Loescher had been asked to work for sixteen hours a day, 'a request far beyond the ability of any man to endure and perform good work',[7] and they decided that no one was to blame for the explosion, clearing him of the accusations.

The accident highlighted the dangers of these boilers. Taking steam from a district heating system seemed like a safer option, as it meant the boilers were placed further away from inhabited buildings. After the accident, more and more people signed up to get steam delivered from the Denver system. Larger pipes were laid underground, insulated with hollowed-out logs lined with asbestos paper, wrapped around the pipe. Metering became standard by the early 1900s, so people were charged for the heat they used as measured by the meter, rather than a flat rate based on the size of the building.

A few years after the Denver district heat system had begun moving steam around the city, a similar system sprang up in London. A district heating network was established using the waste heat from the coal-fired Battersea Power Station, a striking 1930s building that still sits on the River Thames. With its distinctive chimneys, recognisable from the album cover of Pink Floyd's *Animals*, today the building is a collection of luxury apartments and a shopping mall, fittingly with its own central power station and district heating system. Sir Giles Gilbert Scott, the same architect responsible for the classic red British telephone box, designed the exterior of the power station. Between the 1930s and 1980s, coal arrived by boat along the Thames and was tipped into the beautiful art deco building, generating electricity and heat for the surrounding communities.

As with any technology, district heating systems are not perfect. But despite the leaks, communal heat generation does have one big bonus: it makes it easier to change the heat generating source, as only one has to be replaced, rather than many individual boilers or furnaces inside people's homes. This opens up the opportunity for using waste heat or renewable energy technology to generate heat without much disturbance for the end customer: that's you and me at home. While it makes perfect sense to use waste heat, it is not always possible, as power stations and industries that generate waste heat are normally built well out of the way of urban areas, so the heat would have to be transported over long distances to get to the end users. Building long pipes is expensive, and the further the heat travels through these pipes, the more is lost along the way into the environment rather than reaching the end point where it is needed. District heating systems work well in some situations, but they are not the answer in all scenarios.

As a society, we are largely reliant on energy from fossil fuels to heat our built environment, through a variety of technologies like district heating, natural gas boilers, or oil-fired furnaces. The energy-sapping process of generating heat is not going away anytime soon, but I hope and expect to experience a shift away from fossil fuels as the main energy source for heating over the coming years. Some heating technologies will be easier to decarbonise than others – like switching out the heat source for a district heating system – and one key principle will be to make sure any new buildings are efficient, and designed with renewable heating systems rather than relying on fossil fuels.

But heat is only half of the story, affecting those living in colder parts of the world. Many millions live on the opposite side of the coin, where the goal is to stay cool, and many more are set to face the same issue as weather conditions become more extreme and summers get hotter.

*

Of the 2.8 billion people living in the hottest parts of the world, fewer than one in ten have air conditioning.[8] Keeping ourselves cool in heat is just as important as keeping warm, and this is especially a concern for the world's ageing population, as older people are less tolerant of heat than the young; heat can contribute to fluid and electrolyte disorders, renal failure and heat stroke.[9] Keeping cool is not just a comfort, but a necessity for survival – and in a rapidly warming world, it's essential that we find a sustainable way of regulating the temperature of our environments (and by extension our bodies), and that we don't risk exacerbating and extending the problem by burning more fossil fuels to power air conditioners. With so many people living in hot (and increasingly hotter) environments without air conditioning, there is an opportunity to deploy sustainable cooling systems from the get-go, rather than installing something damaging that has to be changed out down the line, after the harm is done.

Although our experience of global warming gives this topic a more urgent edge, cooling is not a new problem; humans have grappled with it through the centuries. The Roman Emperor Elagabalus, who was around in the early 200s CE, allegedly demanded a mountain of snow be built in the garden next to his villa to keep him cool in the summer months. Snow was moved from the nearby mountains by donkey to accommodate this demand, or so the story goes. I can't imagine it lasted long, baking in the Italian sunshine. Nowadays, keeping cool with fans and air-conditioning units powered by electricity accounts for around 10% of global electricity demand.[10] This is of course set to rise as the planet warms up, and unless conscious action is taken, the extra electricity demand will likely be met with fossil-fuelled power generation, keeping us trapped in a vicious circle. But there are some innovative alternatives; it is possible to just skip the middleman – electricity – and instead use a surprising source to cool us down: the sun.

Near Ahmedabad, about 500 kilometres north of Mumbai in India, the sweltering sun beats down on a series of shimmering metallic tubes, attached to the outside of an insulated twenty-foot shipping container. These tubes gather the heat from the sun, but it is not used for heating or generating electricity – amazingly, it is used for cooling. This seemingly magical device is a solar thermal cooling system. Stepping into the shipping container is a welcome break from the heat, which is normally between thirty and forty degrees Celsius in this part of India. But in this ingenious system, the heat from the sun counterintuitively *cools* the container, through a process called 'absorption refrigeration'. Solar thermal cooling systems like this one take the very cause of the heat, the sun, and use it to create a solution for a growing challenge:

the need for sustainable cooling in an increasingly warming world.

'My country requires this sort of solution to reduce our dependency on fossil fuels,' Mahendra Patel, Chairman and Managing Director of the Mamata Group, told me over a video call.[11] The experimental solar thermal cooling shipping container sits just outside one of Mahendra's businesses, a factory that manufactures packaging machines. These test units do not cool anything yet – they are just experiments for now, covered in sensors that have allowed their performance to be monitored for the last couple of years.

The experiment is a novel collaboration between the Indian engineer and businessman and a UK-based cooling company called Solar Polar, which was set up by Robert Edwards and Michael Reid. Robert and Michael met when they happened to sit next to each other at a conference in the UK in the late 2000s. Robert, who had been working on solar collectors for water heating, realised that even in the UK, water can heat up to about 200 degrees Celsius (if the water is under pressure) when it is left sitting still under the concentrated sun's rays. He wondered what could be done with that sort of temperature, and ran the idea of cooling past Michael, a refrigeration engineer. The two engineers decided to spend some time trawling through old academic papers and found patents on solar cooling dating back to the 1800s. But they needed to prove that the concept worked, so they explained the idea to a professor at London South Bank University. The basic numbers stacked up, and they were trusted with a key to the university's labs, under strict instructions to only come in during evenings and weekends, and to put everything back where they found it. By sneaking around the university at night, they were able to test the system and demonstrate that it worked.

As time went on, they continued to develop the Solar Polar cooling system, with help from some unexpected collaborators. 'We worked with the Amish for a long time when we did our early days of development, and they are more knowledgeable than we are,' Robert explained to me when I met him at a café on a rainy day in London.[12] I tried to hide my surprise at this partnership. The Amish are a Christian community known for simple living and slow adoption of modern technology – they prioritise family time and being self-sufficient. Since this cooling technology has been around since the 1800s and can be run off-grid – in keeping with the self-sufficiency desired by Amish communities – some Amish people have been building these refrigeration systems for a long time. One small farming town of 146 people agreed to lend their intuitive knowledge of the technology to Robert and his colleagues.

How it works is simultaneously complicated and simple. Robert told me it took him the best part of a year of working on the system before he could confidently explain how it works; and he still stumbles upon new things from time to time. The key to understanding the thermodynamics of any cooling system is in the evaporation step. Cooling happens when a liquid refrigerant, a substance with the ability to absorb heat from its surroundings, absorbs heat and changes from liquid to gas, or evaporates. Humans do this to cool down. We get hot, we sweat, and the sweat evaporates from our skin, taking heat away with it. In a way, these cooling systems help our buildings to sweat.

In electrically driven cooling, the electrical energy is used to force a liquid refrigerant through a small opening. This leads to a drop in the pressure and temperature of the liquid refrigerant, which turns the liquid to a gas. As this liquid evaporates, it cools the surroundings. An electrically powered compressor then squeezes the gaseous refrigerant, creating a hot, high-pressure gas, which flows through pipes outside of the area being cooled, giving up its heat and turning back into a liquid to start the cycle again. The heat is given up outside the area being cooled – this is why the back of a refrigerator is always warm.

Solar thermal cooling takes concentrated heat from the sun, using it to boil the refrigerant to create the hot, high-pressure gas that gives up its heat to the surroundings and cycles around again. The mechanism of cooling is the same, but the energy needed to run the system comes directly from the heat of the sun rather than from electricity.

For those of you interested in the scientific ins and outs of how this works, it is, by Robert's own admission, complicated. Solar Polar's cooling unit is made up of a closed circuit of interconnected pipes, filled up with ammonia, water and hydrogen. At the start of the cycle, the sun heats up the liquid, boiling the ammonia out of the water at around 200 degrees Celsius. Bubbles of ammonia travel up the pipe, lifting up slugs of water, similar to the way bubbles of boiling water in a coffee percolator move water up through the pot to make a perfect espresso. The ammonia gas flows up, and the water flows down to rejoin the party a little later. Further up the tube, the ammonia gas condenses back into a liquid as it loses some energy, flowing back down a bend into a section of the system filled with hydrogen gas.

When the ammonia hits the hydrogen atmosphere, it behaves as if it has gone into a vacuum and it evaporates – this is where things get exciting. Heat is absorbed, cooling down the surroundings, resulting in a chilly minus thirty-three degrees Celsius. There is now a mixture of hydrogen gas saturated with evaporated ammonia. As this is heavier than hydrogen gas on its own, it starts to flow down, hitting the bottom of the system. At this point, the hydrogen makes its way back up, meeting the water we left behind earlier. The water flows down, passing the hydrogen mixed with ammonia, absorbing the ammonia. We're now back at the start of the cycle, with an ammonia–water mix, and hydrogen gas. And it all repeats itself again. The mixture of liquids and gases stays inside the system for its entire life, going through the same cycle for up to fifty years.

There are no moving parts, so there's no risk of mechanical failure. In the past, impurities coming off the pipes could build up and block the system. Nowadays, cleaner welds and good rust inhibitors stop that build-up of gunk inside the steel pipes. The system is a 'cradle to cradle' design – at the end of its life, the liquid can be drained and the pipes checked, and then the system can either be recharged with new liquid refrigerant ready to work for another fifty years, or melted down for the steel to be reused elsewhere.

Robert spent years developing a trans-Atlantic friendship with the Amish refrigerator makers, and commissioned them to make some units for Solar Polar. I couldn't quite fathom how Robert had managed to find Amish refrigeration experts, but he explained that they do have a website. He also explained that they tend to be very private people, and don't normally talk to 'outsiders', so it took a long time to develop this friendship. 'They are extraordinary people. We went for dinner at the house of the guy who runs the company, him and his sons. We would go there and sit around in a house that is lit by gas light and have the most amazing fruit pie! Everywhere we went, we would be given plates full of slices of pie,' Robert said.

When Solar Polar were ready to use their system in the real world, they wanted to test the prototype in an environment with abundant thermal energy. India was an obvious choice, and Mahendra's engineering know-how and strong interest in renewable energy made him the ideal collaborator. 'When I built my new house in 1996, it was the first residence officially recognised by the government as a fully solar-powered private residence in India,' Mahendra told me. 'I created it out of a commitment to do something new, but it was very expensive at that time.' People called him a fool, apparently, but in fact Mahendra was ahead of his time, and he could see the future in renewable energy before everyone else. Mahendra and his son Manish also experimented with solar thermal air conditioning for the factory, launching the country's first system of its kind in 2006.[13] Tragically, Mahendra lost his son in an accident just a year later, and lost motivation in pursuing this work until he met Solar Polar a few years down the line.

I asked Mahendra why we use electrically driven cooling at all, when using the sun to generate cooling directly seems like a more logical solution. 'The reason is very simple,' he quickly replied. 'It's a matter of convenience. You go into a room and flick a switch, get cooling.' While solar thermal cooling, for the person using it, will not be any more complicated, Mahendra highlighted that it would require a new stand-alone system, whereas the electricity system has been perfected over the years for efficiency, reliability and convenience – as an end user, it is easier to choose the well-known solution your neighbour has instead of the unknown. This is why, as a starting point, Solar Polar will target places without an existing electricity supply, such as rural farming areas where the systems could revolutionise businesses.

'Where farmers grow food, and harvest, they can immediately put it in the cool box, so that they can preserve it until the prices are right and sell it, instead of selling in bulk and not getting a good return,' Mahendra explained. Being able to use cooling on a farm in India could increase income tenfold. Take a dairy farm without refrigeration, where a truck has to come and pick up the milk twice a day. If the truck doesn't show up, the milk goes off. But if the milk is refrigerated on the farm, only one collection is needed per day and the milk stays fresh for longer. Cooling technologies make living conditions comfortable and keep medical supplies, like vaccines, safe to use. But cooling also plays a huge role in reducing food waste. One-third of all food grown worldwide is lost or wasted, partly due to the lack of refrigeration.[14] Not only is this a waste of food, but it's a waste of resources, considering the effort and energy involved in planting, fertilising, watering and harvesting those crops.

The hope for people like Robert and Mahendra is that once the system has proven itself useful for these sorts of applications, people will start to see the benefits and, where appropriate, to choose it over the electrically driven type, saving on electricity use and opening up easier ways to live sustainably on a rapidly warming planet. There is, however, an alternative – or perhaps complementary – solution to all of these systems: designing buildings in a way that reduces the need for heating and cooling in the first place.

*

'That is a roquet, followed by a croquet,' Simon Tilley, a mechanical engineer, said with an air of authority.[15] He had skilfully arranged the six metal hoops in a precise formation on the lawn for my friend Chris's birthday party at his parents' eco-house. Simon, their neighbour, produced a few rolls of thick white twine, deftly unfurling them to mark the borders of the playing area. I stood by the side-lines, precariously leaning over my croquet mallet, in an idyllic setting, with birds chirping and sheep bleating nearby. A three-hour game of croquet ensued. Wooden mallets clanked against plastic balls, punctuated by Simon's encyclopaedic knowledge of the game. Unsurprisingly, Simon and his team claimed the victory, as the sun set over the lush green hills.

We were in the Hockerton Housing project, a community of five eco-homes in Nottinghamshire, in the Midlands of England. I first came across this place a few years ago, when my friend Chris told me his parents had moved to what sounded like a hippy commune. When I arrived, I immediately felt excruciating shame for having driven instead of cycled. A row of strange-looking, hobbit-like homes sat buried in a grass-covered mound, topped with solar panels and surrounded by greenery. A large pond buzzing with insects had a zipline strung across it, and I spotted a couple of small wind turbines in the distance.

The inside of Chris's family home was surprisingly normal; there was nothing 'hippy commune' about it. A paved path led me through a small front garden filled with beautiful flowers, up to the front door. Inside was a kitchen, a living room with a TV, several bedrooms and a bathroom. But I knew these houses were different. I noticed the lack of radiators and spotted the two terracotta pipes running along the ceiling.

The reason we have to heat or cool our homes at all is because heat is constantly being lost or gained through the walls, the floor, the roof, the doors and the windows. The laws of thermodynamics dictate that heat will move from a hot place to a colder place, until an equilibrium is reached. It therefore makes sense to insulate homes, cleverly design them to minimise the heat gain and heat loss, and throw in some heat storage. This will reduce the amount of energy needed for heating and cooling in the first place, and the associated carbon dioxide emissions. This is the general principle behind these innovative Hockerton homes.

Buildings around the world have a different look and feel to them, partly because of the climate as well as the building materials available locally. Our family home in Baghdad had cool concrete floors, and we used to roll out intricately patterned Persian rugs for the winter months to add insulation to the house. In the UK, our house was fitted with carpets all year round, and the windows were double-glazed. When, as a student, I lived near Kuala Lumpur, Malaysia, I was surprised by the open design of my student accommodation block. Stairs, hallways and corridors were open to the outside, to allow natural air flow through the building, cutting through the hot, humid atmosphere. These adaptations in building design do help, but heating or cooling is still needed to maintain a comfortable temperature. The Hockerton homes, however, push design to the limit, eliminating the need for heating and cooling entirely.

'The more insulation you put in, the less heat will flow through it. Which also, on days like today, will keep it cool,' Simon explained on this hot summer's day. Simon and his family have lived in this community of eco-homes for over twenty years. Each member of the community commits one day per week to help with growing food, looking after chickens rescued from battery farms, and maintaining the water system and solar panels, as well as giving tours of the site and doing outreach and educational activities. They reap the benefits through low energy bills, fresh produce, and the satisfaction of doing their bit for the environment.

Each home is an insulated box. The bottom of the box, or the floor, is made up of an insulating layer of polystyrene topped with steel-reinforced concrete blocks. The back wall and the roof, which are buried into a mound of earth, are similar, with a layer of insulation on the outside and concrete blocks on the inside. Burying the building is not necessary for insulation, but the earth roof has other benefits. It can absorb rain, stopping rainwater from gushing off the roof and flooding the ground below. It also makes the solar panels easier to access for cleaning and maintenance, creates more space for growing food, and makes the home aesthetically interesting. The remaining walls are bricks on the outside, a cavity filled with rockwool insulation, and more concrete blocks on the inside. Adding up the dimensions of the different layers comes to just under a metre-thickness of floor, walls and ceiling, which traps heat in the insulating concrete layers away from the living environment during the warm days, and releases that heat during the cooler nights.

Correct orientation of the house to catch the sun is critical, and most normal house builders do not always get this right. In Hockerton, a triple-glazed conservatory sits at the front of the house, and all windows and doors are also made of three layers of glass. The glass lets the sun's warming radiation pass through into the house, but the triple glazing stops heat and coolness from leaking back out. This insulated box gets about half its heat from the sun shining on the house, a quarter from the body heat of the people living there, and another quarter from appliances – cookers, kettles and so on. This creates a problem, as the heat sources are variable. People come in and out of the house, cooking only happens at certain times of the day, and the sun doesn't always shine, so the temperature could fluctuate significantly. But the concrete blocks mitigate this problem by acting as 'thermal mass' – essentially a heat battery, storing the energy.

'It's all down to physics,' Simon told me. 'Heat goes from hot places to cold places, so if the air is warmer, the energy in the air goes into the mass. If the air is cooler, the energy in the mass goes into the air. It doesn't leak out because you've got all the insulation around the outside.' The concrete heat sponge moderates the indoor temperature, maintaining about twenty-two degrees Celsius in the summer and eighteen in the winter. The homes are designed for an average of four people; Simon and his wife Helena noticed the difference when their three children moved out: 'You wouldn't believe how much energy your teenagers can deliver. Metaphorically and physically.'

These eco houses were built with cost in mind; all the materials used can be bought in a standard building merchants, and nothing about them is 'fancy'. Polystyrene, a cheap insulation material made from oil, is a good use of fossil fuel in Simon's opinion, as it lasts hundreds of years and removes the need to spend energy on heating and cooling an entire house. Sealing the house with insulation and triple glazing usually brings up ventilation worries, but for three-quarters of the year, the windows can be opened. For the remainder of the year, the terracotta pipes I spotted in Chris's family home deliver fresh air (via a heat exchanger). Using terracotta, a clay-based ceramic, also reduces plastic use.

Since these homes consume such a small amount of energy, the proportion of energy used for water heating becomes significant, at around a quarter of the overall energy needed. Water is heated in an immersion heater, which is similar to an electric kettle. This makes sense, as the community generates electricity from wind and solar. At one point in our conversation, Simon leaned in as if telling me a secret: 'We're sitting under an experiment.' We both looked up. Above us, sitting on top of a frame, were more solar panels. As these houses are grid-connected, the amount of solar power they are allowed to generate is limited by the utility company, as the cables can only carry so much electricity. This particular system, developed with a graduate from a nearby university, generates electricity independently of the grid, which can be used to heat water.

I asked if they had considered a solar thermal system. 'Solar thermal has water, a pump, and pumps go wrong. Water could freeze. Even when there's only a little bit of sun, the solar panels deliver heat,' Simon replied, echoing Mahendra's point about the convenience of electricity.

Overall, the Hockerton homes use about a tenth of the energy of a conventional house, and it made me wonder why all houses are not designed like this. Simon's theory is that houses sell as they are, so there is no incentive for change. 'People who build houses don't live in the houses, so they have no connection with the heating bill. People who buy houses, who will pay for the heating bill, have no connection with the design of the house. There is no feedback loop, so nothing changes.'

There are other similar housing projects around the world, all tackling different aspects of sustainability. This particular community in Hockerton has cracked sustainable home heating and cooling, and the design could work in other parts of the world, especially those away from the equator, with large variations in temperature. As a species, our need to stay at a comfortable temperature is not going away any time soon, so we have to come up with inventive ways of doing this sustainably, with the least possible contribution to climate change. New buildings, as demonstrated by Simon and Helena's home, can be designed in a way to fit into and work with their environment to reduce or even remove the use of energy for heating and cooling. Designing out the need in the first place is the best option, and should be the first port of call for homes of the future. But the world is not a blank sheet of paper – we are stuck with the buildings and homes that already exist, which still need a source of warmth and an ability to stay cool.

Heating has come a long way from the early days of burning wood, with myriad options in existence, like natural gas boilers, furnaces and heat networks running on waste heat. Looking ahead, there is no one solution to trump all others. Communal heating systems will likely adapt to take heat from new sources, such as electrically driven heat pumps. Existing boilers and furnaces may be modified to run on new low-carbon fuels like hydrogen, all with the goal of reducing carbon dioxide emissions. But the heating and cooling of homes is only one small piece of the energy-use puzzle. Stepping back to look at a fuller picture, the concrete, bricks, glass, steel, rockwool and polystyrene used to construct these homes, and all the contents inside them, still have to be manufactured in the first place, using up energy and resources, and producing greenhouse gas emissions.

When we finished talking, Simon and Helena hopped into the saddles of their tandem bike to cycle off to a protest organised by Extinction Rebellion – a global environmental movement – complete with a banner saying, 'A Healthy World Is Possible'.

'Maybe see you later, if we're not arrested,' they called, waving at me.

Chapter 10

Industry

IN MY MID-TWENTIES, I relocated to Istanbul for six months as the large utility company I worked for was expanding into the Turkish energy market. The company employed more than 10,000 people in many other countries, so our brand-new tiny team of five people, alongside the brand-new location, was a bit of an adventure for me. When I agreed to move, I was excited by the prospect of being in a new place and part of something so unique; but I underestimated how tough it would be to pack up and move to a new country. I didn't speak the language, and I didn't know a single person in Istanbul. But eventually, I found a way to bond with people through a different means; even though many of us didn't share the same language or culture, I was able to form relationships with new people over a shared love of food.

One breezy autumn day, a few months after I had first moved, I found myself hurrying along a jetty to board a boat to take me across the Bosporus, a narrow strait that connects the east and west of Istanbul. I was dreaming of the delights that awaited me on the other side: bright red pomegranates mixed in with deep green parsley, olive oil meandering over the smooth surface of mashed-up chickpeas, deep purple charred aubergines, and simultaneously soft and crispy breads. It had taken me a few long weeks, but eventually I'd managed to befriend a British architect whose company shared the same office floor with mine. Through him, I met a teacher from New Zealand, and a raft of others whom I would get to know over the course of my time in Istanbul. As well as sharing the fact that we were all outsiders, we also ate many meals together, the food giving us a common experience to bond over.

On this particular day, I was on my way to meet my newfound group of friends at Çiya, a homely restaurant that turned seasonal ingredients into delightful, mouth-watering dishes. Çiya's cooks proudly make use of local produce, so most of the ingredients for the meal we ate that day were grown in the country. Blessed with fertile soils, a warm climate and abundant rainfall, Turkey is a haven for growing a huge range of crops, and claims to be one of the few self-sufficient countries in the world in terms of food. About half of the country consists of agricultural land, and nearly a quarter of the population work in agriculture, making the country's agricultural economy among the top in the world. Turkey is the world's top producer of some of my favourite things – apricots and hazelnuts – and they also grow wheat, sugar, tomatoes and other fruits and vegetables.[1] If you hang out in the fruit and vegetable aisles of most British supermarkets and read the labels on the packaging, as I like to do, you will find many of the items were grown in Turkey.

If I've learned anything from my time studying and working in the energy sector, it's that energy is everywhere – it is even inextricably woven into the fabric of every meal you eat.

Without a global energy system, Turkey and other major food producers wouldn't be able to grow and export the food that populations around the world depend upon. Apricots grown in Turkey can travel on a fuel-powered ship to those UK market shelves, while grain from Brazil can travel to China. The people working in the agricultural sector use energy in many different ways to get to the end result of food on a plate – even indirectly. For example, think about the fertilisers and pesticides manufactured in factories that consume energy, which are then sprayed by vehicles that run on fuels like diesel, which also powers the crop harvesters. Without an energy supply for the manufacturing sector to make the pesticides and fertilisers that food growers rely on, the global population would not be where it is today.

In a hypothetical world where everyone ate simple vegetarian diets and farmed the land as efficiently as possible using the best techniques from the 1800s, and the food was successfully distributed and not wasted, the amount produced would only be enough to sustain around four billion people.[2] Today's world population is almost double that number – and rising. Without a doubt, Earth's booming population would never have been possible without some serious changes to the way humans produced food in the last 200 years. And one of the main game-changers that has supported the planet's population of around nine billion is our ability to produce synthetic fertilisers. While fertiliser production may seem like a decidedly unglamorous and unimportant background detail, it is deeply interwoven with energy production and a huge energy user. Any shifts to energy production, and the way we move energy around the globe, will have major consequences for fertiliser manufacture, and therefore the availability of food. We would all feel any food shortages acutely, even in the developed world, where we are accustomed to a reliable flow of goods all year round, whatever the weather and season.

The most important nutrient that plants need to grow is nitrogen, but there is a limited amount of it naturally in the soil. Synthetic fertiliser unlocks this barrier and provides nitrogen, the atoms of which are omnipresent in every protein and the DNA of every cell of every plant and animal. While 80% of the air we breathe is made up of nitrogen, it is locked up in an inert form, unusable by plants, existing as two nitrogen atoms tightly fused together. It's almost impossible to separate these two atoms; they have to be ripped apart in a process that eats up a huge amount of energy, and then transformed into 'fixed' nitrogen before they can become useful for plants.[3]

Without nitrogen, life as we know it is not possible. The copious amounts of nitrogen needed to keep the global population fed is extraordinary, and the production of the synthetic fertiliser needed to deliver the nitrogen to plants is an energy-hungry process. The limited amount of fixed nitrogen in soil gets used up by plants as they grow, and farmers, in response to the demands of a growing population, need ways to replenish it. In the past, people spread horse manure on their fields and rotated their crops, giving the soil a chance to regain its fertility. Populations as far back as the ancient Chinese and Romans recognised the value of doing this, and regularly rotated crops and applied manure fertiliser, but they wouldn't have understood why it worked. Alongside manure, there are certain types of plants, like peas and clovers, that have the ability to provide fixed nitrogen back to soil, and the same is true of certain fungi.[4] These nitrogen-fixing plants are very effective, but growing them reduces the amount of land available for growing edible crops. More dramatically and uncontrollably, lightning fixes nitrogen too; the high temperatures break atmospheric nitrogen molecules apart, freeing them to react with oxygen in the air. The resulting nitrogen oxides dissolve in moisture into nitrates, carried to soils by the rain. However, these natural methods of providing nitrogen to plants simply can't keep up with modern demands.

As the Industrial Revolution gripped the Western world, millions left their agricultural lives and moved to cities. Fewer farmers were left behind to feed a growing number of factory workers. Diets changed too, moving away from vegetables and grains and towards meats, sugars and oils, which require more intensive farming methods. With all these changes and demand for a constant food supply spiking, a way of getting more nitrogen became desperately needed.[5]

The first solution came in the form of guano: bird excrement gathered from rocky Peruvian islands. A step up from horse manure, guano was highly effective, but it was also finite. When that ran out, another natural resource came along to plug the hole: a nitrate mined from the deserts near the Andes in Chile. By the 1900s, Chile produced two thirds of the world's fertiliser. Europe and the United States depended on this, with Germany and Great Britain being two of the biggest buyers.[6] But Chile's nitrate golden age was cut short in the mid-1900s. Unlike guano, this did not happen because of any shortages. A better discovery came along, one that enabled the production of synthetic fertiliser in a way that, at the time, seemed more affordable and sustainable than using excrement or mining a finite resource. This industrial process was so miraculous that it was rumoured it could turn air into bread.[7] But with anything that seems too good to be true, it had a darker side: it could also be used to make explosives.

*

One of the people we have to thank for synthetic fertilisers, and the accompanying horrors, was a German chemist called Fritz Haber. Haber's interest in chemistry flourished early – his childhood saw him carrying out experiments in his bedroom, until his father had enough of the smells coming from under his door and banned his son's hobby.[8] He was born in 1868 into a thriving Jewish community, at a time when Jews were barred from some professions, but many respected positions were within reach. Perhaps to get around this issue, aged twenty-four, Haber converted to Christianity. It was more as a formality than out of any genuine religious feeling – his true god, after all, was science. Haber focused his efforts on being German, and his national pride and science became increasingly intertwined, guiding his life.[9] After securing a well-respected position as a university professor, Haber got married and had a son, but throughout his personal life ran a common thread of self-centeredness and distance. His wife, Clara, who was the first woman to earn a PhD in chemistry at her university, once wrote, 'Fritz is so scattered, if I didn't bring to him his son every once in a while, he wouldn't even know that he was a father.'[10]

Looking back on the timeline of his career, I suspect Haber neglected his family in part because he was completely consumed by the ammonia problem – his work was focused on trying to persuade nitrogen in the atmosphere to react with hydrogen to form ammonia, which could then be turned into a fertiliser. It was a pressing issue for humanity at the time, and many chemists would have felt the pressure to solve it and claim the prizes of prestige, respect and money waiting on the other side. Making ammonia – combining a nitrogen atom with three hydrogen atoms – is a tricky chemical reaction. The reaction consumes large amounts of energy, and it also gives off energy as heat, burning up the freshly made ammonia. Haber and his colleagues found ways to overcome these issues. They tried a more suitable catalyst, a substance that pushes the chemical reaction forward without getting used up, and ran the reaction at a higher pressure. As a result, the temperature could be reduced, and less ammonia burned away.[11] At the time, Germany was dangerously dependent on Chilean nitrates, so in 1908, Fritz received handsome research funding from a large chemicals company, BASF. The following year, Haber had a breakthrough. He gathered his colleagues around, and they watched as less than a quarter of a teaspoon of ammonia dripped into a flask.[12]

This was a turning point in history; Haber had found the answer to one of the biggest scientific challenges in the world. But it was a tiny machine, making a tiny amount of ammonia.[13] Scaling up the process from a quarter of a teaspoon to enough ammonia to produce fertiliser for the world took a few more years, and input from another BASF chemist, Carl Bosch. The BASF executives tasked Bosch with looking into

Haber's machine, and the two men worked together to build a bigger one. As the size went up, the problems grew, but Bosch was not fazed. He had grown up around invention and engineering; his father ran an engineering business and his uncle was Robert Bosch, founder of the German appliance maker. Unlike his colleagues, who dressed up for work and only spent time with other scientific staff, Bosch spent his time with the company's labourers and personally worked on the machines.[14] His experience helped him to guide his team and develop many world firsts as they scaled up ammonia making: the biggest compressors, the strongest valves, new materials for the reactors, and large quantities of hydrogen and nitrogen gases.[15]

The barrage of challenges was handled methodically and meticulously by Bosch and his team, who found ways to extract hydrogen from coal and nitrogen from the air in large quantities to use as the raw ingredients needed to make ammonia. By 1911, the machine produced a few tonnes of ammonia a day at a cost cheaper than anything else on the market. From here, BASF went on to build multiple full-scale Haber-Bosch synthetic nitrogen factories, and Bosch eventually became the head of the company.[16] Both men even received the Nobel Prize for Chemistry for their work on ammonia: Fritz Haber in 1918 and Carl Bosch in 1931.

But there is a dark side to making fertilisers – chemically, they are alarmingly close to explosives. The same discovery that could feed the world could also destroy it. Governments want to be able to make the chemicals for a reliable fertiliser supply to feed their population, but also to be able to manufacture munitions for their armies. Following the start of the

First World War in 1914, the German government made a deal with BASF to make the chemicals needed for explosives. Bosch was torn; he did not want his beloved machine to be used for destruction, and he referred to the deal as 'this dirty business' during the negotiations. Despite this, by May 1915, Germany was making the chemicals at home after being cut off from Chilean nitrates by the Allies.[17]

Haber, who had moved to a new academic position in Berlin by this point, was full of scientific ideas that could help Germany win the war.[18] He became a top adviser to the Kaiser, respected and welcomed, and acted as an intermediary between the government and BASF. In 1918, a new factory was built in the centre of Germany, close to coal mines – for energy as well as the raw material needed to make hydrogen – and close to sources of water. The factory could be used to make fertiliser in peacetime, and explosives during wartime.[19] Some historians have even estimated that the war would have ended a year or two sooner without the Haber-Bosch process.[20]

That same nationalistic pride that Haber had nourished early on in his life soon saw him making other cold-blooded scientific contributions to the war effort, including using chlorine gas as a chemical weapon. This earned him an Iron Cross, but came at the cost of his marriage.[21] His wife Clara was horrified that her husband was turning his brilliance to such terrible ends. She was a talented chemist but had increasingly been expected to fulfil the role of a traditional German wife, devoted to *kinder, küche, kirche* – kids, kitchen and church. She found life suffocating, writing to a friend that 'what Fritz has achieved ... I have lost, and even more'.[22]

Using chemistry to kill others was the last straw for her. Clara went to the garden, taking her husband's revolver with her. Hermann, their son, ran to the garden after hearing two shots. He called to his father, but it was too late by the time Fritz got there. Clara was gone.[23]

The end of the First World War did not bring peace for everyone. Bosch and others spent the years that followed working hard to change international perceptions of German businesses. As the years went by, Bosch was dismayed to see the growing popularity of Hitler and the Nazis. Not only was Germany headed for ruin, but this was also bad for business.[24] In January 1933, Bosch's worries escalated when Hitler was named Chancellor. A few months later, a decree to cleanse the government of non-Aryans was put in place. By this time, Haber was in his fifties and battling declining health. His conversion to Christianity and obsessive dedication to his country counted for nothing.[25] He was filled with self-hatred and felt like his whole life had been a lie. 'I was German to an extent that I feel fully only now, and I'm filled with incredible disgust,' he wrote to a friend.[26] Motivated by the idea of getting British citizenship, he moved to Cambridge in England to take up a new academic post: 'My most important goals in life are that I not die as a German citizen,' he wrote. The goal was not achieved; he died in January 1934, still a German.[27] Perhaps this was a blessing in disguise, as he did not live long enough to learn that his research into chemical insecticides would later be developed into Zyklon B, a poison gas used by the Nazis to kill millions of people in concentration camps.[28]

After the end of the Second World War, the factories that had been manufacturing weapons went back to being used for Fritz Haber and Carl Bosch's original intention: to make ammonia for fertiliser to feed people.[29] Haber and Bosch's lives, filled with success and struggle, reflect the conflicting story of synthetic fertiliser production. That conflict continues to this day, not just because of the association with weapons and war, nor the associated greenhouse gas emissions from ammonia production – there are other environmental issues at play, too. On the one hand, about half of the nitrogen in our bodies comes from a Haber-Bosch factory – the synthetic nitrogen is now in our DNA, sustaining life in a positive way.[30] But on the other, inefficient and excessive use of fertiliser means that much of the fixed nitrogen ends up washed into waterways or in the air, rather than in food. These nitrates flow into the oceans, upsetting the natural chemical balance and impacting the creatures living in the water. Unlike the days when it was a struggle getting hold of enough fertiliser, the ease of its availability today has enabled large-scale monoculture farming – where only one type of crop is grown at one time on a specific field – and the vast quantities of grains grown enable factory-farmed animals.[31] Not only is this a cruel practice, but it can create a hotbed of diseases that get passed onto wildlife and negatively affect humans.[32]

Alongside the recognition and correction of the environmental issues caused by synthetic fertiliser production, we will need to decarbonise the ammonia production process if we want to have any hope of maintaining a healthy food supply for a continually growing population. Decades may have passed since Haber and Bosch gave the world fertiliser on tap, but producers still follow the same fundamental recipe and chemical reaction steps from all those years ago. It takes far less energy to make a tonne of ammonia today than it did back then, but manufacturing ammonia globally still uses up nearly 2% of the world's energy output each year. Most of that energy is used to supply a steady stream of hydrogen for the process, and the most common way of making the hydrogen is by splitting it out from hydrocarbons, which spews carbon dioxide into the atmosphere. This is also part of the reason that ammonia production is responsible for around 2% of global greenhouse gas emissions.[33]

The carbon dioxide that is produced during the making of ammonia is normally blown off into the atmosphere, but it is possible to capture it and use it in fizzy drinks, or as a cooling fluid in certain types of nuclear power stations. Capturing the carbon and using it elsewhere is one way to temporarily reduce emissions, but swapping out the fossil fuels for renewable sources is the real key to reducing emissions in a meaningful way. Unlike electricity production, the fossil fuels are not only used to generate electricity, but to provide the raw materials, hydrogen in this case, to make the fertiliser. In ammonia production, about 40% of the fuel is used as a feedstock – or an ingredient – which turns into ammonia and eventually fertiliser, so it is key to provide these ammonia production facilities with low-carbon hydrogen.[34]

Food is obviously an essential component for life, but it comes at a cost to the environment. Our population faces a dilemma – we have to sustain everyone on the planet with enough fertilisers, but we cannot continue using them as we have in the past while ignoring the consequences. It is in everyone's interest to minimise the use of fertilisers to reduce the negative impacts on the planet. China, the world's largest producer of ammonia, accounting for 30% of all production,[35] alongside other large producers in North America, Europe, India and the Middle East, is very aware of this issue, and these areas have started to at least think about or even trial low-carbon ways of producing ammonia. The road to decarbonisation looks promising, but very long – shifting a decades-old industry from the status quo will take some concerted effort and perseverance, but it is our only option if we want to reduce emissions from the industries that sit behind the scenes, propping up our lives.

*

As well as being delightful for my tastebuds, Istanbul is an enchanting experience on the eyes. My first Saturday in the city was set to be lonely, but I happened to find someone in a similar situation – my forty-two-year-old South Korean neighbour, who was also new to the city and appeared to feel as lost as I did. Despite the lack of a shared language, he managed to communicate to me that he had been sent to Turkey by the bank he worked for, and we were able to spend the day exploring, appreciating the sights and sounds of the city. Over the coming weeks and months, I spent a lot of time walking the streets, from old cobbled alleys to narrow staircases wedged in between buildings, and busy bridges crossing water. I admired Istanbul's exquisite buildings: the grand Dolmabahçe Palace, the Hagia Sophia Mosque, the Galata Tower. The Basilica Cistern, hidden from view underground but an equally impressive structure, is a dimly lit cavern lined with fat stone columns. The space was formerly used to store water.

But built around all these ancient sites is the connective tissue of our modern era: roads and pavements made of asphalt, buildings constructed with modern-day concrete, steel and glass. After working in the energy industry for so long, when I look at the buildings and structures around me, I see megawatt hours. I see energy consumption. The link between buildings and energy is similar to that between food and energy, in that the webs of energy spun underneath the surface remain unseen. For buildings, most of the materials have been extracted from the earth, like iron ore for making steel, which is processed in a factory, heated and turned into the right shape, moved to the building site and assembled. One of the most important and ubiquitous building materials, not just in the modern era, is concrete, which is made up of a mixture of sand, gravel and crushed-up rock glued together with water and cement. Second only to water, concrete is the most consumed material in the world.[36]

It's no exaggeration to say concrete is everywhere – as is the cement embedded within that very concrete. Making cement is the most energy- and carbon-intensive process of the manufacture of concrete; it's responsible for about 8% of the world's carbon dioxide emissions, accounting for the vast majority of the emissions associated with concrete manufacture.[37, 38, 39] If the cement-producing industry was a country, it would be the third largest emitter in the world, behind China and the US.

The impacts of this industry stretch beyond beautiful structures and sightseeing; the mixture of sand, gravel, rocks, water and cement is in everything around us. Buildings, roads and bridges are the obvious applications, but it pops up in huge quantities in the construction of nuclear power stations, wind turbine foundations, and any hydroelectric power station, like Cruachan in Scotland or the Three Gorges Dam in China. Concrete is versatile; it can be poured and formed into any shape. It is affordable, easy to make and has the right structural qualities for durable buildings. It's also a safe material – it doesn't catch fire, and it can be reused at the end of its life for other applications like building roads. Like it or not, concrete and cement are part of your life; you indirectly use energy by walking along a road, driving across a bridge, or even just being in an office building with concrete foundations. As an individual, it would be hugely impractical to reduce your own emissions by completely avoiding concrete – it is not as easy as switching to a renewable electricity supplier or hopping on a bike instead of driving a car down the road. It will be down to the people in the business of making the stuff, and governments, to push and enable a change. However, an awareness of how the industry works will help individuals to make informed political voting choices, or to make an educated judgement about how far along certain sectors are with reducing emissions.

My awareness of this industry developed a bit later in my career, when I worked for the UK government, running energy innovation competitions. The premise was to give funding to projects and companies working on energy technologies that will help reduce greenhouse gas emissions. Some of the funding went to cement manufacturers who were trying out low-carbon fuels to power their factories.

That was how I first met one of the people working hard to change the industry – Dr Diana Casey, executive director of energy and climate change at the Mineral Products Association, a UK-based trade association for cement and concrete, as well as other mineral materials.[40] A career in grey powder and crushed-up rocks may not be a typical dream job, but there's more than meets the eye when it comes to concrete.

'It has been around for a really long time, so we know how to use it,' Diana enthused when we spoke recently, before excitedly adding that the modern-day version of cement, concrete's gluey counterpart, celebrates its 200th birthday in 2024. As if to remind me of its ever-presence, at exactly that moment a concrete mixer, with its familiar slowly rotating drum, appeared at a building site visible through the window of the room we were sitting in. Two hundred years is a long time, but it is thought that much earlier iterations of cement-like binders existed. Almost 3,000 years ago, the Nabatean Bedouin tribes, in today's southern Syria and northern Jordan, used it to build underground water cisterns, very much like the Basilica Cistern I saw in Istanbul.[41]

But how is cement physically made, and why is it so carbon-intensive? I decided to see it all for myself, and arranged a visit to a cement factory (another glamorous location on a grey, wintery day). My tour started in a quarry, where excavators were removing huge chunks of limestone rock, before the mechanical teeth of the crushers chewed them up into smaller pieces. The crushed-up rock then made its way through a series of tubes and conveyors, before being dumped into a kiln – a giant, hot, rotating horizontal tube. Standing a metre or so away from the kiln, I could feel an uncomfortable heat radiating off it. In the kiln, a flame heats everything up to over 1,000 degrees Celsius. The flame comes from burning coal and other fuels, and the heat splits the limestone into calcium oxide and carbon dioxide. Marble-sized hard grey balls called clinker come out of the other end. Once the clinker cools down, it is ground up into a dust and mixed with other materials – a little bit of gypsum and ash – to give the end product the right properties.[42] Finally, in a mesmerising mechanised packaging room, fine grey cement powder drops into plastic bags, which get sealed and labelled. The process is repeated over and over again with an almost hypnotic rhythm, although only a small amount of the powder ends up in bags – the rest is moved in bulk to concrete factories by road tankers.[43]

At a global level, about 40% of the carbon dioxide emissions from cement-making come from burning fossil fuels to heat the kilns. Another 10% come from the fuels used to provide the energy for quarrying, moving and grinding raw materials, cooling and mixing – this is fuel for vehicles or electricity to run machines. The remaining 50% of the emissions come from the chemistry of the process – when limestone is cooked, it breaks down and releases carbon dioxide.[44]

The energy-related half of the emissions can be tackled by making the process more efficient, so that it needs less energy, and by switching to low-carbon fuels. Most cement makers rely on coal, but over the last few decades there has been a switch away from coal to some rather creative fuels. 'Across the sector, there are now about fourteen different fuels used,' Diana told me, listing waste tyres and any combustible materials that we put in the rubbish. Other biomass fuels are used, too, like waste wood, sewage pellets and meat and bonemeal – the stuff left over from an animal carcass. Switching fuels is no easy task, though, and on the whole, the cement industry's energy use still has some way to go to become sustainable – globally, about 90% of the fuel used is still coal.[45]

The emissions from the chemistry half of the process are more difficult to deal with. That fundamental chemical reaction, which releases so much carbon dioxide, has to happen to make the cement product we have been using for the past 200 years – there is no other way around it. The most viable solution at the moment is carbon capture and storage: catching the carbon dioxide that is going up the chimney from the factory, and storing it away from the environment – in an empty underground oil field, for example. Carbon capture and storage already exists, but not at a large scale. There are even designs out there for fossil-fuelled power stations fitted with carbon-capture technology. 'I should say that carbon capture is going to be quite energy-intensive itself,' Diana warned. I personally see carbon capture as a last resort, to mop up any remaining carbon dioxide emissions we cannot get rid of any other way, like process emissions from cement – it is not an alternative to fixing the emissions problem at the root.

If carbon capture is added to a cement plant, Diana and the Mineral Products Association argue that the factory should still switch to lower-carbon biological fuels rather than relying on carbon capture to catch the emissions from coal, reminding me that 'if you are regrowing the biomass, then you have that potential for negative emissions', meaning that the crops harvested to provide the biomass are replanted. In theory, this is the best way to use biomass fuels. However, the amount of biomass fuels is limited by the available land to grow them, so it makes sense to only use these fuels where they are absolutely needed, and to capture the carbon dioxide emissions wherever possible.

Another possibility to explore is a natural property of concrete called carbonation, which is where a concrete building breathes in carbon dioxide from its surroundings during its lifetime, offsetting the emissions released during manufacture. Several groups are now working to increase this property, for example by crushing concrete at the end of its life and spreading it out to create a larger surface area that will absorb more carbon dioxide.[46]

Looking ahead, especially for built-up countries with stable populations, I would have thought that minimal amounts of concrete would be needed, because the majority of the infrastructure – houses, roads, bridges – already exists. Surely using less concrete in the first place has to be the best option here? Diana told me that, yes, there is a big difference between the concrete needed in an urbanised place like the UK compared to booming countries like India, but 'we still seem to need new houses ... and infrastructure projects like HS2 and Thames Tideway,' she said, referring to a new UK trainline and a super sewer project in London. On top of that, many of the energy infrastructure projects themselves also need concrete; in 2019, the UK's Hinkley Point C nuclear power station set the record for the longest continuous concrete pour, which went on for five days.[47]

You may wonder, is there not a better material out there? Have we not invented anything else in the last 200 years? Can we reduce concrete use by partially replacing it with other, less carbon-intensive building materials? One alternative is steel, but that comes with its own very similar energy-intensive production process and carbon dioxide emissions. There is some scope for building with timber, but it is difficult to imagine a timber skyscraper, and it does not quite work for wind turbine foundations and other big infrastructure projects. As Diana summarised, 'If you can have net zero concrete, why would you keep cutting down trees? We need our trees in the ground.' Aside from its carbon emissions, concrete is unquestionably incredible – so incredible that most of us take the safety and security of the structures we inhabit for granted. Any alternative building material will have a mountain to climb to gain our trust and prove itself as a worthy replacement.

Concrete and cement are here to stay for the foreseeable future, even for built-up places, so decarbonising production is a must. The UK is relatively advanced on this journey compared to some others – coal makes up only about half of the fuel used in cement manufacture, with trials and plans ongoing to continue reduction. Other solutions, like carbon capture and storage, are being trialled in Belgium, Canada and the US. As with ammonia production, there is definitely a shift in the sector. Encouragingly, this 200-year-old industry is willing and committed to change on paper, but it will take time and patience for the infrastructure needed to enable that change to come into play.

*

Will turning off fossil fuels and replacing with renewables solve all of this? Unfortunately, not quite. Industry uses up about 40% of global energy supplies, in the form of electricity, high-temperature heat or to make the feedstock (ingredients needed for making something). And herein lies the problem: simply switching to renewable electricity will not get rid of all the emissions. Clean electricity is not a replacement for feedstocks.[48] All these manufactured materials, the literal building blocks of our society, suck up energy in the process of becoming a product – and energy has a financial and environmental cost, even if it is clean.

So instead of narrowing our thinking to maintaining the status quo, there are other angles to tackle this from – one is to reduce overconsumption and waste. Wasting food, by overeating or allowing it to rot, isn't just food waste – it is a waste of all the energy that went into making the fertilisers and pesticides used to grow the food, the fuel used in the machines that ploughed the fields, and the fuel used in the trucks that moved the food from field to plate. The same applies to buying an item of clothing and wearing it once, or electronics, or any other product.

Plastic, a relatively new invention, in theory helps to reduce food waste by enabling its storage and preservation. But as we all know, single-use plastics lead to a huge waste of resources, not to mention the damage done to the natural environment and oceans from the plastics that end up in there. Before the invention of synthetic plastics, people used natural materials with similar properties, like animal horns, tortoiseshell, amber or shellac, to make everything from cutlery to combs and billiard balls. Of course, as demand went up, these animals suffered. The invention of synthetic plastic may have saved them, but the downsides are being realised now.

What is the future solution to all of this? It is not straightforward, but a balance of different avenues is needed. First and foremost, reducing waste and overconsumption can go a long way to bringing down the amount of energy needed to produce these products. Every single individual can play a part in this, by being conscious of what is behind every object – the convenience and ease of our world comes at a cost, and a shift in attitude is needed from the throwaway culture that exists today. But it's about more than individuals; it is a shift in culture that needs to come from the people working for and running governments and businesses.

Reducing overconsumption will not entirely eliminate a product, so the enormous opportunity to improve collection of waste and recycling across the globe is another viable way to reduce the need to make new things from scratch. And then there is the search for alternative raw materials; in the case of ammonia and fertiliser manufacture, producing hydrogen by electrolysis powered by renewable electricity is a cleaner alternative to extracting hydrogen from coal or natural gas.

When it comes to synthetic materials like plastics made from fossil fuels, one option is to use biological raw materials. Waste from the food and drinks industry can be turned into the raw materials needed to make the chemicals that go on to become products. Another option is to use renewable electricity to capture carbon dioxide from an industrial process or from the atmosphere, and chemically combine it with clean hydrogen to make a synthetic hydrocarbon that can form the basic raw material for many products.[49] Clean technological options do exist for the transformation of the chemicals industry, but this is in the early stages, and completely dependent on the shift of the energy sector from fossil fuels to renewable energy.

Cement, concrete, ammonia, fertiliser and many other industries exist hidden from view, using up energy. It can be difficult to draw a direct line between industry and an individual, but manufacturing only exists to serve us – we are indirectly using that energy when we have a meal or chat away on a work call in an office. Every concrete and steel building, every item on a supermarket shelf – it's all come directly or indirectly from one of those factories somewhere across the world, made possible by another heavy energy user, the transport sector, which moves us and our goods by cars, trucks, trains, ships and planes. It would be unreasonable to place the responsibility of industrial emissions on individuals for using offices and shopping in supermarkets; it is almost impossible for us to function outside of the existing system. But by opening our eyes individually to the processes behind all of these things that we rely on every day, we can start to at least understand and tackle, piece by piece, the energy use and associated negative consequences from the industrial world that makes life so much easier for many of us.

Chapter 11

Transport

I REMEMBER FEELING as if I had been in the car for an eternity. It was dark by now and I had run out of questions for my parents, songs to annoy my siblings with, and games to play in my head. So, I grabbed a jumper and carefully placed it against the inside of the car door, creating a comfortable nest for my head. I closed my eyes and lay back, but something compelled me to open them again. What I saw simultaneously sent warm shivers down my spine and brought tears of amazement to my eyes. For a moment, I forgot to breathe. Thousands and thousands of stars, pinpricks of bright white light, flooded my vision and blew my tiny seven-year-old brain. This exact image of the Milky Way is still imprinted in my memory, along with the feeling of awe.

We were driving through the North African desert, along the coast, somewhere between Tunisia and Benghazi in Libya. The petrol-powered car we were in was transporting us from one country to the other. We couldn't take a plane, because of air sanctions on Libya, so driving was the only way there from Tunisia. A car may seem like just a car – a mundane lump of metal that takes a person to work every morning or home from an evening out. But to me, and countless others, it is much more than that. During the first years of my life, cars helped me and my family move to safer countries. Aged nineteen, a car rushed me to the Emergency Room after I fell, hitting my face on frozen ground and spilling blood onto the hard, white ice. A year later, a car took me into Malaysia's Taman Negara National Park, where I witnessed another breathtaking light show, this time of fireflies.

These vehicles have been reliably there for me, through difficult times and moments of sheer joy. But as I grew older and moved to a city, we have grown apart. I am now surrounded by bicycles, buses, overground and underground trains, boats cruising along the river, and planes up in the sky above me. I don't own a car and feel good about the carbon emissions I am saving. But I live in a city with an excellent public transport network, and I'd be kidding myself if I thought that I wasn't still hugely reliant on fossil fuel-powered transportation – we all are. Whether or not you own a car, there are still trucks, trains, passenger planes and ships moving us around, and the goods we all rely on, like food, medicines and electronic gadgets, which are manufactured in one corner of the world and shifted over to another.

Transport around the globe is one of the heaviest users of energy, accounting for about a third of all energy consumption.[1, 2] Within the transport sector across land, sea and the skies, it is sea transport, largely for shipping, that uses up the smallest proportion of energy. Air transport, primarily for passengers, uses slightly more than sea transport. But it is road transport that is the biggest culprit, accounting for around three-quarters of the total energy consumption of the transport sector in 2019.[3,4] This extremely high use of fossil fuels in the transport sector accounts for a fifth of global energy-related emissions, and it bathes us in copious amounts of air pollution spewed out into the atmosphere.[5]

Given that transport generally, and road transport in particular, accounts for so much environmental harm, it seems like an obvious place to start as engineers, scientists, politicians and members of the public begin to think about what actions will have the highest impact in combatting climate change. One obvious yet extreme solution is to dramatically cut the movement of people and goods around the world. But staying globally connected is important; I have hugely benefited, personally and professionally, from being able to travel and get to know other cultures. As we saw earlier, Henry Stimson, the US's Secretary of War, chose not to drop a nuclear bomb on Kyoto because he had visited the city in person and liked it. The benefits of being closer together cannot be underestimated – ideas cross borders, furthering scientific knowledge, and travel widens our circles of friends, colleagues and partners across the world. I would not want to take this privilege away from anyone – this is why switching up the energy use of land vehicles, planes and ships is so critical.

Over millennia, we have figured out how to cover greater distances at higher speeds. New territories were explored and more goods were moved between regions. From simply walking, we began using domesticated animals for transport, before devising the wheeled vehicle to attach to those animals. But the innovation that made the biggest difference to society and brought widescale long-distance transport to the masses – as well as contributing to the climate crisis we are facing – is the now-ubiquitous automobile.

*

One early morning in 1888 in south-west Germany, a woman and her two teenage sons set off on a unique journey. They climbed into a three-wheeled contraption with a combustion engine, and drove from Mannheim, where they lived, to Pforzheim, some hundred kilometres away, to visit family. The woman was Bertha Benz, the wife of Karl Benz, the German engineer credited with inventing the first practical automobile.

Karl Benz had patented the first vehicle powered by a gas engine two years prior to the journey, as early as 1886, and Bertha herself had been heavily involved in the invention. A visionary individual, Bertha came from a wealthy family. Despite protestations from her father, she invested her dowry in her unsuccessful and penniless future husband, and continuously pushed him to trust in his abilities and persevere through the hard times.[6] At the time of their marriage in 1872, Karl was drowning in his failing iron construction company, but Bertha's investment dragged him out of this and funded some of his future work. Not only that, but she also worked alongside him in the workshop and was integral to his success.

The couple had a permit from the Mannheim district office to drive their novelty vehicle around the area, but Bertha wanted to push it further.[7] When her husband woke up that morning, he found a note on the table saying something along the lines of, 'We're off to see my parents.' He assumed she had gone by train, until he realised that the patented car was missing.

Since cars didn't yet exist, there was no road network like today's, nor any navigation systems. Bertha drove alongside streams and railway lines to find her way to Pforzheim, going at a top speed of sixteen kilometres per hour – a good cycling speed today. But the engine was not particularly powerful at 0.75 horsepower, so the car could not go uphill. This may be where the two teenage sons came in useful; when they encountered inclines, they would push the vehicle uphill while she steered the car.

Bertha faced and solved multiple problems along the way. She used her long hairpin to unblock the carburettor and a garter to insulate an electrical wire, and she got a shoemaker to replace the worn-out leather straps of the brakes. Without any fuel stations, she had to work out her own way of getting petrol. A pharmacy in Wiesloch provided the goods; she picked up a small amount of ligroin, a petroleum product used as a solvent that could double up as fuel. She had to stop at several other pharmacies to buy enough fuel to keep the vehicle going.

Bertha and the two boys got to Pforzheim after dusk, sending a message to Karl by telegram to let him know they had arrived. After a few days there, they made their way back in the vehicle. Bertha's genius was two-fold; she used the journey to identify kinks in the design and suggest technical improvements to Karl, including adding one more gear for going up inclines, but she also used it as a remarkable advertising opportunity. People who saw them along the way were perplexed by this 'horseless carriage', and just as she intended, the trip received a lot of publicity. The car's first foray into the public eye went a long way towards allaying any fears by showing it was a safe way to travel. In 1889, Karl Benz had the confidence to show off their invention at the World Fair in Paris, and their cars began to sell.

What started as a clunky three-wheeled carriage evolved into the cars that have changed lives all over the world today, my own life included. And the journey that started it all, Bertha's adventurous off-road route, has been turned into the scenic Bertha Benz Memorial Route, running through south-west Germany.[8, 9] From those humble beginnings, the number of cars has expanded beyond what Bertha and Karl Benz could have ever imagined, spurred on by the availability of cheap oil to propel them. The number of motor vehicles swelled as the oil and gas industry expanded in the twentieth century. Around the time of Bertha's death in 1944, there were roughly sixty-five million cars on the road, most of them in the US.[10] These numbers pale in comparison to the decade we are currently living in, with over 1.4 billion cars on the planet that enable us to have different choices, meet different people and see different places.

It is unrealistic to expect 1.4 billion trusty and useful cars, and all their emissions, to disappear overnight. Rather than a magical disappearing act, there is a transformation creeping in. As of 2020, the number of electric cars on the road reached ten million.[11] While this is only a tiny fraction of the total number of cars, the rate at which this fraction is growing is rapid. I have noticed this change happening around me. In recent years, friends and colleagues have silently rolled up in their electric vehicles to give me a ride somewhere. The novelty of being in an electric car is beginning to wear off, marking its entrance to normal life.

Quietly cruising along, electric cars give off a misleadingly futuristic vibe. They have actually been around since the 1800s – in fact, by 1900 in the US, they accounted for a third of all vehicles on the road. They were great for short trips around cities, and a popular choice for women in particular, as they required less brute force to drive than the gasoline-powered car. Electric cars also came without the noise and exhaust fumes, but they could not quite compete with fossil-fuelled combustion-engine cars as that technology improved and those cars got cheaper. Today, electric cars are a viable option again, as we find ourselves at a climate change crossroads.[12]

On the surface, electric vehicles may seem like the ultimate solution for reducing emissions from driving, but they are only as emissions-free as the energy source they use up. As the proportion of electric vehicles goes up, more renewable electricity generation will be needed to charge them, along with extra cables to transport that electricity. But one major plus is that these cars could give back by being mini batteries on wheels – helping with the balance of electricity supply and demand. While plugged in, electric vehicles could suck up electricity when there is too much generation, or act like tiny batteries and give electricity back to the grid system if there is a deficit. Most cars are only driven for a couple of hours a day; the rest of the time, they sit idle on a driveway or in a car park, with plenty of time to spare to help the electricity grid system – the system operator can charge or discharge the vehicle's battery, while making sure there is enough juice in the 'tank' for the owner. As the number of fossil-powered cars drops, and electric vehicle ownership rises, the lines between electricity generation and use will become more intertwined and blurred. Rather than being just a user of energy, like today's petrol and diesel cars, vehicles of the future could play multiple roles in the energy system – enabled by digital technologies and more automation.

But it is not quite a done deal: electricity will not solve all our transport emissions. The larger the vehicle is, like a double-decker bus or a heavy goods vehicle, the bigger the battery needed – becoming impractically heavy in some cases. Some medium-sized vehicles have managed to go electric; about 60,000 of these were sold worldwide in 2022, and China sold around 50,000 new electric buses, driven largely by the need to reduce air pollution as well as greenhouse gas emissions.[13]

Another option could be to fuel vehicles with low-carbon hydrogen, negating the battery weight issue, or simply as an alternative option to electric if that need ever arises. Hydrogen passenger vehicles have been around for a while; the world's first mass-produced version was the rather sleek-looking Toyota Mirai, gliding into markets in 2014.[14] But since then, hydrogen-powered cars number only in the tens of thousands[15] (paling in comparison to electric cars), roaming streets mainly in South Korea, Japan and China.

I often see these two technologies pitted against each other, 'electric versus hydrogen', but it doesn't have to be a battle; both technologies come with their own pros and cons, and could play a part in the energy transition. One benefit for electric cars is the charging infrastructure. It is more advanced at the moment, while hydrogen refuelling stations lag behind.

Having a car that you can't refuel is clearly a major hurdle. On the other hand, many drivers worry about the range of electric cars – will it have enough juice for a long trip? Will they have to add significant time to the journey for recharging the batteries? Hydrogen cars tend to offer greater range, and faster refuelling times.

For heavier modes of transport, like the heavy goods vehicles that shuttle all sorts of things around for us, hydrogen could be a practical option. Perhaps if hydrogen refuelling stations are built for heavy goods vehicles, hydrogen passenger cars would tailgate closely behind. In a clean energy future where people continue to heavily depend on energy-consuming cars, buses, trucks and trains to get to work, meet new people from across cultures, spend valuable time with friends and relatives, and get their daily necessities, there will be room for electric, hydrogen, and perhaps other renewably fuelled vehicles.

*

One summer, I travelled for a site visit with some colleagues to Campbeltown, a small community in a far-flung corner of west Scotland, set around a harbour facing out to sea. The first issue we had to solve was how to get to Campbeltown – in distance it's not that far, but in terms of transport it is relatively inaccessible. From London, the obvious way is to fly the 550 kilometres to Glasgow, followed by a short flight in a small plane out to the peninsula the town is perched on. But when you are in the business of clean energy and decarbonisation, taking the most energy- and carbon-intensive form of transport feels counterproductive. I convinced my colleagues that we should take the Caledonian Sleeper – a 'sleeper' train that traverses the distance overnight.

I won't lie, it was not the best night's sleep I have ever had, but our journey emitted about 80% less carbon dioxide than the equivalent flights, and I think I could eventually get used to sleeping on a moving train. While it is not always possible, I do consciously look for other options to travel instead of flying. Clearly, this is something many of us are becoming more conscious and careful of. European rail companies are expanding night trains to give passengers the option to go to sleep in one country and wake up in another; the Austrian Federal Railways are adding routes to Barcelona, Paris and Amsterdam.[16] Taking a domestic flight emits more than three times the amount of carbon dioxide compared to taking a train or a fossil-fuelled car filled with four passengers,[17] but in some instances flying is unavoidable, and often it is quicker or simply more convenient. Driving or taking the train is not always practical or possible, so just as cars are shifting fuels, so too are planes.

Some of the younger, climate-conscious generation are driving this major shift in the aviation industry. I met one of them, Robyn Huiting, a twenty-two-year-old industrial design student at the Eindhoven University of Technology in the south of the Netherlands.[18] Robyn was thrown in at the deep end when she joined as the manager of the Falcon Electric Aviation team. The student team was set up in 2019 to make existing small aircraft more sustainable. Rather than running on fossil fuels, the team wanted to develop electric power train conversion kits to allow the existing planes to switch to renewable electricity – this is akin to removing someone's major internal organs and putting in a whole new system to keep them alive, while keeping the outside the same.[19]

Comfortable in her 'design bubble', as she described it to me when we spoke, Robyn combines her creative and technical interests by studying the aesthetics of products and how users interact with them, as well as the technical side of how to physically make a product. She came across the Falcon Electric Aviation team and was drawn in by the future-focused nature of the project, which extends beyond her life right now. She wanted to see how 'die-hard' engineers work, and she also wanted to 'work on something bigger than just a more basic energy topic', she told me over a video call from Denmark, where she was studying at the time as part of an exchange programme.

This super-young team of engineers has proved that the concept of converting an existing plane from fossil fuels to electricity is viable, and they hope to complete test flights and gain regulatory approvals over the next few years. They fit in this work on top of their studies, driven by a passion for aviation and a desire to contribute to the solutions for climate change. To prove the concept, the team were able to take apart and put back together a two-seater Cessna 150 plane that now runs on a battery, which can be charged with electricity from renewables, rather than on fossil fuel. A two-seater plane is obviously very small, but as batteries develop and improve, it may be possible to apply this conversion concept to larger aircraft. 'The conversion part is the same,' Robyn explained, 'the only difference is the battery.' At the moment, there is a limitation; as with road vehicles, the larger batteries get, the heavier they are, making them impractical for larger planes. As a starting point, the Falcon Electric Aviation team are designing a packaged conversion kit, so that the conversion can be done by aircraft manufacturers in their own factories.

Robyn is at the beginning of her life as an adult, so I was curious about her views on transport and energy use. While it is not something she thinks about all the time, she said she is willing to pay a little more to go by train rather than flying because of the impacts on the climate. Among her friends and peers, Robyn told me that 'conversations about this are frequent, so it makes you think about it more'. Looking back at my own thoughts and attitudes at that stage of my life, choosing anything but the cheapest option would not have even crossed my mind. Twenty-two-year-old me had very little knowledge and awareness of the energy consumption and emissions associated with different modes of transport, so I feel hopeful when I see that younger generations are much more conscious of energy use when it comes to travel.

Knowledge and awareness are part of the challenge, but they are not enough on their own, particularly when our choices are constrained by other factors, such as affordability. A large cost problem still exists – the cost of a train ticket compared to a flight ticket does not reflect the energy consumption or impact on the environment, so making the right environmental choice is not always possible, let alone easy. As part of her exchange programme to the university in Denmark, Robyn was offered €50 to travel by train instead of other modes of transport, but that was not enough of a saving to justify not going by car or plane. 'It's a nice start and it's good that they think about it, but if you don't have the money, you still go for the easy way,' Robyn said.

Not satisfied with adapting older machinery, many organisations and start-ups are working on the design and testing of new electrically powered aircraft. Designing a new product from scratch is easier in some ways, as there are fewer restrictions to deal with compared to converting an existing product. But for Robyn and the team, reusing existing aircraft is a big part of sustainability, otherwise products get left behind in the transition to clean energy. Robyn explained that, as she got deeper into the aviation project, the importance of thinking about the full life of a product became clearer. 'I'm not sure in the beginning I understood that was the purpose,' she told me, but now it has impacted her viewpoint when it comes to product design – thinking not only about a new product but how to avoid waste at the end of its life.

Conversion to electricity may work for small airplanes, but it is a different story for larger ones. Long-haul flights, like those I have taken between the UK and Canada, Malaysia and Turkey, mean aircraft are in the air for many hours, transporting hundreds of people. An immense amount of energy is needed to keep these planes in the air, so batteries are not currently an option, as they would be far too big and heavy. For comparison, the two-seater Cessna 150 burns through about twenty litres of fuel per hour,[20] whereas a larger plane like a Boeing 747, which can hold about 500 passengers, gets through 13,000 litres per hour.[21] One alternative is to switch to 'Sustainable Aviation Fuel', which is chemically the same as a hydrocarbon fossil fuel, but a manufactured version made either from biological materials like plants or wastes, or by combining low-carbon hydrogen with carbon dioxide captured from an industrial process like steel- or cement-making. At the moment, only small amounts of Sustainable Aviation Fuels are manufactured and used globally – but efforts are intensifying as nations wake up to the effects of climate change and the reality that we need to change our energy consumption habits in every aspect of our lives. In the longer term, low-carbon hydrogen could be used directly as an aviation fuel, removing the steps needed to combine it with other molecules to make Sustainable Aviation Fuel. The likes of Airbus – a large multinational aerospace company – are currently designing and testing aircraft engines to run on hydrogen.[22]

I may feel good about not owning a car, but how my life has unfolded, and some of the choices I have made over the years, make flying unavoidable. Many members of my extended family also left Iraq and landed in different corners of the world. My husband and his family are deeply rooted in Nova Scotia, Canada, a flight across the ocean from the UK. My work, like the energy system itself, is not bound by country borders and requires me to travel, often by plane. I am closely watching the aviation sector, hoping my flights will switch to using a different fuel and emit fewer greenhouse gases as soon as possible, so I can start to feel better about flying and let go of the guilt in favour of the joy it brings to my life.

*

On a stormy night in January 1992, the rough waters of the Pacific Ocean were so violent that huge waves knocked a number of shipping containers off a cargo ship. One of the forty-foot containers split open, releasing 28,800 rubber bath toys into the sea.[23] Thousands of yellow ducks, green frogs, red beavers and blue turtles, made in Chinese factories, had been destined for American bathtubs but never made it. Drifting with the whims of the waves, they toured the world, passing the site where the *Titanic* sank, and landing as far afield as Japan, Alaska and Hawaii. Some unfortunate plastic floaties spent years frozen in an Arctic ice pack.[24] At the time, the shipping company didn't own up to the spill and the origin of the toys remained a mystery until a journalist, Donovan Hohn, investigated, documenting his findings in his book *Moby-Duck: The True Story of 28,800 Bath Toys Lost at Sea*.[25]

Around 90% of the world's cargo travels by sea, packed up inside shipping containers – standard-sized metal boxes that can be easily moved between ship, train and truck. Millions of these metal boxes are moving across the globe on ships at this very moment. A few, like the floating bath toys, do end up in the sea. While this can often be cause for alarm and problematic for marine life, when it came to the bath toy spill, scientists made the best of a bad situation by mapping the movement of the rubber floaties, and using the information to track ocean currents. They have also tracked the movements of 60,000 Nike shoes, 34,000 hockey gloves, and five million Lego pieces that have all been accidentally released into the oceans.

The vast majority of shipping containers and their contents do reach the intended destination, but that would not be possible without an energy supply to move the ships, giving them a ride across the world. The large ships tend to run on heavy fuel oil, a substance left over from the crude oil refining process after kerosene, jet fuel, gasoline and diesel – which smaller ships and boats use – have been taken out. The heavy fuel oil left looks like tar; it is so thick that it has to be heated before it flows well enough to be burned by a ship's engine. As you may expect, it is nasty if let loose on the environment.

All manner of stuff gets moved in ships, such as food grown using ammonia fertilisers, shoes, toys and building materials. In 2019, a whopping sixty-five million tonnes of iron ore extracted from Australian mines was exported to Japanese steelmakers, with the ships burning heavy fuel oil to move the weight between the two countries. In the same year, all of the ships traversing the three major East–West containership routes moved around twenty-four million shipping containers, burning about eleven million tonnes of fuel. Added up, global shipping emits about 3% of total carbon dioxide emissions[26] – more than Germany's annual emissions alone.[27] Yet, just as with personal transport, cutting back on the transport of goods is not a realistic option for reducing emissions. Japanese steelmakers need the Australian ore to make steel for the country's infrastructure projects, and Western Europe relies on corn, wheat and barley grown in Ukraine and moved by ship across the Black Sea.[28]

Some areas are better placed to grow or make certain things because of their climate, access to a natural resource, or specific skills the population is trained up in, so exchanging goods with each other makes sense. Moving all this stuff by ship is the best option from an emissions perspective compared to trucks or planes, but there is still huge room for improvement. This is what drives Michael Vahs, today a professor of ship operations at the University of Applied Sciences Emden/Leer in Germany, who previously worked as a ship's captain on many types of vessel, including cargo ships.[29] Since 2000, when he started his university position, Michael has been researching how to make shipping greener and more sustainable. When I first met Michael over a video call, he was surrounded by models of boats and boat parts in his office near Hamburg, and his eyes glittered behind his seafoam-green-rimmed glasses as he told me about his love for shipping. 'If I wasn't from a shipping family, I would have taken this choice anyway,' he said, explaining that his father and grandfather were also involved in shipping. 'I feel attracted by the coast and by the sea. This is where I like to live and be.'

Before speaking with Michael, when thinking about greener shipping, my mind immediately jumped to the development of sustainable fuels to replace fossil fuels, similar to those being developed for aviation. But Michael and his colleagues are reviving an old and effective method of moving ships by directly using the wind with sail systems. This may sound like an antiquated, backwards step if you are imagining Viking longboats or the fleet of the Spanish Armada. But the future of sails could be very different to the past. 'The field is growing, and we are now running many projects where we bring sails back to the ships,' Michael told me, adding that other means of propulsion are also needed alongside the sails to meet today's tight shipping schedules.

Before the development of steam and diesel engines, people relied heavily on the wind to get them across the oceans. But travelling aboard a steam-powered vessel was comfortable and reliable, and plentiful cheap coal and subsequently oil were available. 'We adjusted to this. In the beginning of the fossil age, we never thought that we were going to run into such a big problem by using those resources,' Michael reflected, comparing the climate change problem to a ship that is running full speed into a rock, with a short amount of time to correct its route.

In the lab, Michael and his colleagues build small-scale models and test them in the wind tunnel or water tank, transferring the results to full-scale applications. While some of the work is theoretical, a lot of it involves real life. In 2014, Michael got a phone call from the University of the South Pacific, a university owned by the governments of twelve Pacific Island countries with its main campus in Suva, Fiji. 'Are you the guy who is researching sail systems?' the voice over the phone asked. The call was just the start of what is now a pioneering collaboration between universities, governments and shipping companies to develop sustainable shipping.

Taking part in this collaboration are the Marshall Islands, a collection of paradise islands and atolls surrounded by turquoise lagoons. But like their neighbouring island nations, they are exceptionally vulnerable to climate change and rising sea levels.[30] As such, politicians there have prioritised climate policy, well ahead of other nations. For a country made up of islands, shipping is vital for moving goods and people around, so for the Marshall Islands, finding a way to reduce emissions from shipping is an imperative. Michael and his colleagues across the various organisations formed a project to develop small boats that are as close as possible to carbon-neutral for local lagoon transportation, and larger vessels for transportation between the islands. Working with the Marshall Islands Shipping Corporation, a state-owned shipping company, the team assessed the local needs and came up with something robust and suitable for such a remote location – where getting a spare part to repair a fault on a vessel can take a very long time.

During the project, Michael visited the Marshall Islands several times. He described the place as special and unique, full of palm trees and coral reefs, with a friendly and respectful culture. So friendly, in fact, that the education minister offered Michael Marshallese citizenship at the beginning of the project. 'They have a good sense of humour,' Michael told me.

The larger cargo vessels move goods from the main island, Majuro, to the outer islands. Copra – a dried coconut product – is returned from the outer islands to coconut oil production facilities in Majuro. The existing vessels run on diesel; the new design is similar but with a high-performance textile sail system and solar panels to generate electricity for lighting and living areas. On board, there is room for eight overnight passengers, and space for additional apprentices to learn about low-carbon shipping operations. The ship's propellor can also operate as a turbine to generate electricity when there is good wind. Including a small diesel combustion engine in the design is inescapable to give reliability when there is no wind, but one of Michael's inventive students has shown that biodiesel made from coconut could be used as a replacement for diesel. When I spoke to Michael, the ship was about to be constructed in a shipyard and would be put into service around a couple of years later, with him aboard – he hoped.

The new vessels could be game-changing for island transportation. While this is a step in the right direction, it is sadly not comparable to huge cargo ships or the large cruiseliners people board for tourism. These goliaths may not be popular with younger generations, as 'they are more aware of what they are doing and their carbon footprint', Michael suggested. With big cruise ships, Michael feels we have lost sight of the point of tourism – to see different places, experience new cultures and relax. This can hardly be achieved if 4,000 people simultaneously descend from a cruise ship into a picturesque port town for a short period of time. 'Downsizing might be a trend,' Michael speculated, with consideration being given to sustainability as well as affordability. This is what he hopes his students and the upcoming generation of young entrepreneurs – like Robyn and her engineering colleagues – will achieve.

This may be the case for cruise ships, but what about goods transportation? All those items of food, yellow rubber ducks, Nike shoes, hockey gloves and so on are moved about on colossal ships, which would be nearly impossible to power with wind alone because of their size and weight, coupled with the impatient demands of the modern world. Alongside directly using the wind to move ships, Michael told me about some other solutions he has come across in his work, one of which is surprisingly simple: going slower. The relationship between the speed of a ship and the amount of fuel it uses is not linear, meaning that going twice as fast does not consume twice as much fuel – instead, the relationship is cubic. So, if a vessel travels twice as fast, at least eight times the power is required; hence, reducing speed can drive down energy demand with very little effort. This seemingly easy step can make a huge difference, but perhaps it does not fit into our impatient, fast-moving, consumption-focused culture. Changing this culture, if at all possible, would be a very slow process, but reconsidering existing shipping patterns with this solution in mind could uncover some routes where ships can slow down and save fuel.

Another important measure is to increase efficiency, including checking if a vessel has the optimum parts, down to the smallest details, like using the right type of paint on the hull to reduce friction. This can be time-consuming and hard work for an efficiency gain of a few percentage points. Once the efficiency options are exhausted, that leaves biofuels, liquid hydrogen, or new sustainable, synthetic hydrogen-based fuels made using renewable electricity. Fuels like methanol and ammonia are energy-intensive to manufacture, and even if you are using renewable sources of energy, there will still be energy losses along the chain of manufacture, from the wind or solar farm generating electricity, to conversions to hydrogen and then ammonia, and all the way to the losses when the fuel is consumed by the ship's engine. Those few percentage points gained by improved efficiency, from things like the ship's paint type, become ever more important.

'This is why we believe sails systems are good – they use wind power directly. We believe this is a good building block of zero emissions shipping in the future,' Michael told me proudly. He added that this renaissance of sail systems is slowly entering the consciousness of the shipping industry, and pilot projects have resulted in ships being retrofitted so around a quarter of their power comes directly from wind energy. 'This is significant, if you can solve 25% of the problem,' Michael explained. But the technology side is only one dimension of decarbonising shipping and beyond. Work is needed in all areas – in politics, the media and society itself, as well as in technology. Michael believes higher levels of ambition, empathy, and a common approach that puts ego aside is needed: 'Then we will be sailing into a world of justice, and not a world of chaos.'

Just like Bertha Benz, Michael and Robyn are willing to push the boundaries to develop new technologies to improve global transport systems. The benefits of travel were demonstrated yet again through my conversations with them – Michael has been fortunate enough to spend time in the Marshall Islands and get to know the local communities, and Robyn is gaining valuable experiences from studying abroad, just as I did.

Over the years, the pace and distance of travel has ballooned as humans have learned to find and harness energy supplies, to our detriment in many ways, but to great benefit too. It is too late to put that genie back in the bottle – we have experienced the wonders of travel, from transporting us to safety, or bringing us food and resources from across the globe, to enabling friends and families to spend valuable time together. It's all possible because of energy resources being manipulated to propel these machines across land, sky and sea.

Speaking to Michael and Robyn gives me hope. But it is not just them – on my journey in the energy industry, I have encountered many others dedicating their lives to building sustainable energy systems. While the technologies exist, they are not perfect, but they will still move us a step in the right direction, and this is better than doing nothing and waiting for a miracle. It is time to transition away from fossil fuels, and in parallel actively seek improvements for society, and remain ready and flexible enough to adapt to moving along the correct path towards a future beyond the climate crisis.

Epilogue

FOR A PERIOD in the mid-1990s, my siblings and I lived with my mother in Amman, Jordan, while my father worked in the UK. We anxiously waited for two long years for permission from the UK government to join him. My mother even overheard my four-year-old brother discussing the visa to the UK with his little school friends – a rather grown-up conversation for such young children to be having. During this time, I regularly hand-wrote and posted letters to my father, and he would call us on a landline telephone about once a week, as international phone calls were still so expensive. We were eventually reunited, building a new life in the UK in a quaint English village, with reliable energy supplies and less temperamental cars than the Volkswagen Passat 'Barazilli' we'd had in Baghdad. For many years after moving to the UK, my parents had limited contact with their families in Iraq and across other parts of the world – until the internet came along.

I was a teenager when we got our first computer and painfully slow dial-up internet; connecting to the internet produced a tinny and very distinctive dial-tone sound – anyone who was around during this time will have this etched into their memory. Being connected to the internet blocked up the phone line – you had to choose between being on the internet or having an available phone line for people to call (it feels like a lifetime ago, before the days of smartphones). I spent hours chatting to school friends online, attempting to send MP3 files of songs we had recently discovered to each other, only for the internet to disconnect randomly, undoing the hours of progress of sending the song – I would then reconnect, and start again.

Today, like many of us, I am constantly connected to the internet through a number of devices, and rely on these for directions to get me around, online banking, listening to new music, looking up answers to questions that spring to mind, and, of course, keeping in touch with people. It is hard to imagine going back to life in the early 2000s and before. As someone with a terrible sense of direction and no memory for maps, I have no idea how I would get anywhere without my smartphone. Instead of just a few phone calls a year, my parents can now speak to and even see their friends and family across the world anytime they want through video calls. Through social media, they have the latest family news, gossip and pictures on tap. Undoubtedly, the vast connective web that is the internet has changed many people's lives for the better, but like everything, it comes at an energy cost.

Currently, there are around five billion internet users worldwide, consuming vast amounts of energy needed to run the infrastructure.[1] It is tricky to put a number on the energy demand, partly because it depends on how you define the boundaries of where the demand begins and ends, and also partly because it is a moving target as different devices and users connect or disconnect. Energy is consumed by data centres densely packed with servers that store and process data; infrastructure that transmits information from the data centres to users, employing routers, switches, communications towers, and so on; and of course the laptops, smartphones, fitness trackers, pet cameras, and any other internet-connected device. Estimates do vary wildly (with some studies disagreeing by five orders of magnitude), but one estimate from 2020 puts the number at around 5% of global electricity use. Given that electricity only makes up about one-fifth of the world's total energy use, that puts the energy burden at around 1% of our global energy use – arguably a tiny share for the benefits delivered.[2]

As with other 'stuff', like food, fertiliser, cement and glass, the energy consumption of the internet stretches far back into the supply chain. All the devices we use – as well as the cables and equipment that make up data centres – are made from materials that have been mined and manufactured. Until we fully transition to a renewable energy system, energy consumption equals greenhouse gas emissions – so the existence of the internet and all that it has brought us, good and bad, also pollutes the physical world.

The impacts extend beyond getting, moving and using energy to power the internet. Today's latest gadgets quickly turn into tomorrow's forgotten toys, so what happens to your old phone or laptop when you upgrade to the new model? At the moment, probably not much. While there are some electronic waste recycling schemes, the global rates are inadequate at best. In the vast majority of cases, all the effort and energy that went into mining the metals, manufacturing the plastic and carefully constructing the circuit boards inside the devices ends up in the wasteland of a landfill. As more people access the internet and become reliant on more connected devices, electronic waste will pile up beneath our feet. There has to be a better way: keeping our devices for longer, having the ability to fix them rather than replacing them to be brand new, and developing better recycling facilities. I hope these will become the norm in the near future.

But I worry about existing initiatives that push for mining the deep oceans or asteroids for the precious metals needed to make our devices; what new environmental damage are we willing to cause? Surely, we have the awareness and know-how to better manage the existing resources on our own planet? Expanding mining feels very much like someone taking out multiple loans and credit cards, fully knowing they will not be able to pay them off. If we continue along this path, we will increase our waste generation to levels that spiral out of control, and as a society, we will eventually have to declare bankruptcy.

The internet and associated digital technologies could be viewed as a burden on society, polluting and using up large amounts of electricity and resources, but this pales in comparison to the energy consumed by our transport and manufacturing sectors. Connected devices can also save energy; think about the travel emissions that are avoided by switching to virtual meetings. And with increased computational resources and data availability, we have an opportunity to see the issues related to climate change and the energy transition under a bright new light – these digital technologies, enabled by the internet, can be a positive force for change.

EPILOGUE

Every step of the energy journey – extracting energy resources, converting energy from one form into another, moving it around the globe, and the multitude of ways that energy is used – can generate enormous datasets. Measuring and analysing these data, building up an overall connected picture, and modelling different future scenarios will be key to making decisions about what direction to take.

However, while modelling is helpful, it is far from perfect. Our world is complex, and energy systems do not exist in isolation, so gathering data and building digital models has its limitations and is bound to miss the nuances of culture, human behaviour, and many other immeasurable factors that play a role in how we do things. Perhaps this is where Artificial Intelligence will step in – unpacking the complexity present throughout the energy journey in new ways. According to one definition, 'Artificial Intelligence is a machine's ability to perform the cognitive functions we usually associate with human minds.'[3]

In today's world, a small team of smart people could get together and put their human minds to work by using historical data to build a model or simulation for a particular part of the energy system, such as a natural gas pipeline network, and then run new scenarios – what does it look like if we take out the natural gas and replace it with hydrogen? What happens if these pipelines are hit by extreme weather events? How does that impact carbon emissions? Do the pipelines deliver the energy needed by the users at the other end? These models give us a best guess, but they are only as good as the information that goes in, and the model assumptions made. They fundamentally rely on the minds of that small group of people, so it is likely they will miss something, despite their best efforts. As AI develops, machines will increasingly be able to learn, reason, interact, problem-solve and even be creative – a trait we generally view as deeply human. For the energy system and climate change, AI may be able to come up with ideas we had not previously thought of or solutions we had not known how to implement.

Thinking about all of this in the context of energy throws up limitless opportunities. Take global transport systems: improvements to the overall efficiency of how people and things move across land, sea and the sky would reduce the energy demand, edging us a step in the right direction on greenhouse gas emissions. At a local level, AI could piece together data to offer a driver the most fuel-efficient route, based on up-to-date road conditions and weather information. Meanwhile, the timing of traffic light changes across a city could be optimised to improve transport flows. But if everyone is driving electric cars, why does efficiency matter? Lowering energy demand ultimately means fewer wind turbines or solar panels needed, shorter cables to move electricity, less concrete consumed and materials mined, and less of our environment taken up by energy infrastructure.

AI will continue to revolutionise how our products travel across the world. Just as drivers can take fuel-efficient routes, a ship's captain can sail a more efficient route across international seas. Instead of relying on the experience of a few people, data-driven decisions can inform how fast a ship should travel and when it needs be maintained or even cleaned to reduce its fuel use. For the final leg of a product's journey, software can now help to arrange parcels inside a delivery truck, like a game of Tetris, to minimise empty space. Yet again, this reduces the overall energy needed to get a product from the maker to the user.[4]

These are some of the applications we have already come up with; the future is bound to bring even more imaginative solutions. AI and the connectivity provided by the internet can help us to pinpoint and fix the energy inefficiencies in our society. How can we mine materials like iron ore or rare earth metals, needed for devices and energy sector equipment, in a more efficient way? What is the optimal route for a high-voltage cable or a subsea pipeline, avoiding environmentally sensitive areas? How do we make sure every rooftop solar panel is being utilised in the best way, supplying the owners with electricity or storing or exporting it to an electricity distribution network at the right time? Adjacent to the energy transition, AI and connectivity could also help us to improve the recycling of electronic waste – hopefully eliminating the need to mine our oceans and outer space.

Rather than being a drain on the energy system, entertaining people with cute cat videos or generating harmful deep fakes, these digital technologies can have a positive disruptive effect on the energy system. Examples of this exist already; in 2019, Google DeepMind developed a system to predict the output from wind farms thirty-six hours ahead of actual generation, using weather forecasts and historical wind turbine data. This improved the wind farm operator's ability to commit to hourly delivery of electricity to the power grid, helping them with balancing electricity supply and demand ahead of time with renewable power.[5] In a similar vein, an AI-based system developed in the UK was 33% more accurate at forecasting output from solar.⁶ With these systems, the Achilles heel of renewables – the intermittency – suddenly becomes more predictable and therefore manageable.

As with any technology, it's not all rosy. As we've already seen, there is of course the energy demand and carbon footprint associated with computing and AI – these deep learning systems consume large amounts of energy during the training phase, so the carbon benefits need to outweigh the cost. Socially, there is a huge risk of exacerbating inequality between the rich and the poor across the globe – today's AI development is predominantly led by academic institutions and companies in wealthier countries, where it would be easy to overlook the needs and experiences of the rest of the population, many of whom are at most risk from the impacts of climate change.⁷ Underpinning any AI development with a strong code of ethics, and expanding the diversity of the industry, will be critical to avoid this pitfall. It is also dangerous to relax and believe that AI will fix all our climate problems for us – we need to remain very much in the driving seat.

But, of course, all of this relies on us making the right choices. Had we made different choices around the time of the Industrial Revolution, we would be in a completely different world right now. We are at another critical juncture in the history of humanity. I don't think there is one correct version of a sustainable energy system, no blueprint we can pull out and implement; there are many different options, each with pros and cons. I would be very (pleasantly) surprised if whatever we do next turns out to be perfect, and we can sit back and enjoy a glorious, affordable, clean and sustainable energy system.

EPILOGUE

What we do next has to reduce the harmful impacts of climate change, but we need to be aware that our next steps may create new, unforeseen problems. But this is the cost of progress – it is no reason to sit back and do nothing, waiting for a miracle.

While we do all live under one sun, each region has its own natural resources and challenges. Australians have to survive extreme heat, but the same sunny heat is a valuable resource, while Norwegians deal with bitterly cold winters, yet have plenty of flowing water for hydropower. Each geography will have different options, but there is a delicate balance to be struck. After all, we are all tethered together by the energy infrastructure that runs between countries and continents, and increasingly so by digital technologies. Each region has to be aware of what its neighbours are doing and manage the interface seamlessly.

We have to make changes fast. That applies to those of us in the energy sector, but whatever your calling is, you can play a role. You can stay informed and make environmentally conscious business decisions and consumer choices where possible. You can also engage with decision makers and politicians, be that writing to a local representative or simply turning up to vote for a political party committed to doing something about climate change.

The changes we will make over the coming years will be incremental, using the best knowledge that we have, so we must be ready to adapt and adjust in the future. We can't spend long years over-analysing the options to make a decision, while the world literally burns around us. There are many great tools in our toolbox, and many talented engineers, scientists and other energy sector professionals – it's time to use the tools and focus talents. Wherever we go next, we have to respect the interconnectedness of the planet and our energy system, and work with nature rather than fight it.

Further Reading

WRITING THIS BOOK gave me good reason to seek out and read other books about energy. I am eternally grateful to the authors who dedicated time to researching and telling the stories of energy, and I truly enjoyed the books listed below. I would recommend them to anyone looking for a deeper dive into these topics.

- *Sustainable Energy – Without the Hot Air*, David MacKay, UIT Cambridge, 2009
- *Energy: All That Matters*, Paul Younger, Teach Yourself, 2012
- *Atomic Awakening: A New Look at the History and Future of Nuclear Power*, James Mahaffey, Pegasus Books, 2009
- *Let It Shine: The 6,000-Year Story of Solar Energy*, John Perlin, New World Library, 2013
- *The Hydrogen Revolution: A Blueprint for the Future of Clean Energy*, Marco Alverà, Hodder & Stoughton, 2021
- *The Electric War: Edison, Tesla, Westinghouse, and the Race to Light the World*, Mike Winchell, Henry Holt & Company, 2019
- *The Grid: The Fraying Wires Between Americans and Our Energy Future*, Gretchen Bakke, Bloomsbury USA, 2016

- *The Alchemy of Air: A Jewish Genius, a Doomed Tycoon, and the Scientific Discovery That Fed the World but Fueled the Rise of Hitler*, Thomas Hager, Broadway Books, 2008

Acknowledgements

THE IDEA FOR this book has been brewing in my head for close to ten years now. Knowing full well that I could not make it happen alone, I often closed my eyes and thought about getting to the point where I would express my deep gratitude to everyone involved in this journey. I am finally at that moment, which feels surreal, so here goes . . . First of all, thank you to my wonderful editor Anna Baty, who saw something in my ideas and made them even better with her suggestions, and to Izzy Everington for picking up the baton and patiently working through chapters with me. Another big thank you to everyone else at Hodder and Stoughton who was involved in this process, including those of you I did not get to meet. Also, thank you to my agent Elizabeth Sheinkman, who took a gamble on an engineer with a desire to write a book.

Next, a big thank you to my interviewees; these valuable members of the global energy community took time out to share their experiences with me: Pasi Tuohimaa, Melanie Windridge, Amit Singla, Rahul Gupta, Greg Morrison, Tatjana Pletena, Sarah Kimpton, Michele Nesbit, Cara MacEachern, Trevor Schulz, Robert Edwards, Mahendra Patel, Diana Casey, Peter Roberts, Michael Vahs and Robyn Huiting. Another giant thank you to those who not only spoke to me, but openly showed me around their worlds: Fiona Macleod at Lynemouth power station in England, Britta Jensen and Allan Jensen at Tvindkraft in Denmark, Hanne Tvedt and her colleagues at METCentre in Norway, Ian Kinnaird and Sarah Cameron at Cruachan power station in Scotland, and Simon Tilley at the Hockerton Housing Project in England.

To all the members past and present of Neuwrite, thank you for not only critiquing my early drafts, but sharing your collective years of science writing experience and showing me the myriad creative ways in which science and engineering can be communicated in writing; I hope I can make you all proud. Alongside Neuwrite, thank you to everyone who helped me in the early days, in many different ways, to get this book from an idea to the real thing. There are far too many people to list who gave me advice, introduced me to others, or helped and influenced in some way; but to name just a few: Vince Pizzoni, Mark Miodownik, Roma Agrawal, Alex O'Brien, Manisha Morais, Malcolm Bambling, Ian Wilson, Sean Atton, Elizabeth Hillier and Adam Duckett.

After all the interviews, research, writing and editing was done, huge doubts crept into my mind – will any other humans actually enjoy reading any of this? Did I interpret all this information correctly? Have I made thousands of mistakes? So, a massive thank you to my interviewees, who reviewed what I had written, and to my friends and colleagues, who spent their free time reading chapters and giving feedback: Roy Allan, Rose Durcan, Phil Rogers, Mike Miller, David Brownjohn, Jim Briggs, Henning Wehn, Kerrine Bryan, Tom Haslam, Paul Guillon, Lauren Rice, Laura Koster and

ACKNOWLEDGEMENTS

James McGrane. An extra-special thank you to Beatriz Pizarro-Aparicio for being a fountain of emotional wisdom, and to the incredible three people who read the entire book: Ian Llewellyn, Ken Wright and Amber Hosking.

Another thank you to everyone in the global energy sector – you keep our world going. You warmly welcomed me into the industry, and since then have given me a meaningful career and collaborative atmosphere to thrive in – I am excited to play my part in the energy transition, and optimistic that together we will move in the right direction.

Finally, the hardest thank you to express in words is for my family. Ronald, thank you for not only patiently listening, but actually engaging in my endless book chat over the past few years. Thanks for being my number-one cheerleader, giving me unwavering support, and filling my life with humour and laughter. To my incredibly talented sister and brother, Nadine and Jahfer: thank you not only for reading chapters, but for being there through the past (and future) ups and downs of life. And to my mother and father, I probably don't say this enough: I love you both and I am eternally grateful for everything you do for us.

Endnotes

Introduction

1 Hannah Ritchie, Max Roser and Pablo Rosado, 'CO_2 and Greenhouse Gas Emissions', Our World In Data, accessed 26 August 2023, https://ourworldindata.org/co2-and-greenhouse-gas-emissions
2 'Statistical Review of World Energy, 2023', Energy Institute, accessed 26 August 2023, https://www.energyinst.org/statistical-review

Chapter 1

1 Richard Rhodes, *Energy: A Human History* (New York: Simon & Schuster, 2018)
2 John Evelyn, *A Character of England as it was lately presented in a letter to a noble man of France* (London: Joseph Crooke, 1659)
3 'Great Smog of London', Britannica, accessed 5 February 2023, https://www.britannica.com/event/Great-Smog-of-London
4 Vaclav Smil, 'Energy in the Twentieth Century: Resources, Conversions, Costs, Uses, and Consequences', *Annual Review of Energy and Environment*, 25 (November 2000): 21–51
5 Author visit to Port of Tyne and Lynemouth power station, and interview with Fiona Macleod, 29 March 2022
6 'Bioenergy', IEA, accessed 1 October 2023, https://www.iea.org/energy-system/renewables/bioenergy
7 Becky Mawhood, Dominic Carver, Sarah Coe, Paul Bolton, Alex Adcock, 'Sustainability of Burning Trees for Energy Generation in the UK', House of Commons Library, 29 November 2022, https://

researchbriefings.files.parliament.uk/documents/CDP-2022-0220/CDP-2022-0220.pdf
8 'Indirect Land Use Change (ILUC)', European Commission, 17 October 2012, https://ec.europa.eu/commission/presscorner/detail/en/MEMO_12_787
9 'Global Coal Plant Tracker', Global Energy Monitor, accessed 5 November 2022, https://globalenergymonitor.org/projects/global-coal-plant-tracker/
10 'Global Energy Review 2020 – Coal', International Energy Agency, accessed 5 February 2023, https://www.iea.org/reports/global-energy-review-2020/coal
11 Diodorus Siculus, *Library of History Volume I* (London: Loeb Classical Library, 1933), https://penelope.uchicago.edu/Thayer/e/roman/texts/diodorus_siculus/2a*.html
12 Darrin Qualman, 'Happy motoring: Global automobile production 1900 to 2016', Darrin Qualman, 13 June 2017, https://www.darrinqualman.com/global-automobile-production/
13 As of 2022
14 John Chipman, 'Iraqi Farouk al-Kasim behind Norway oil fund that is envy of world', CBC, 13 April 2014, https://www.cbc.ca/news/canada/iraqi-farouk-al-kasim-behind-norway-oil-fund-that-is-envy-of-world-1.2604105

Chapter 2

1 James Mahaffey, *Atomic Awakening* (New York: Pegasus Books, 2009), e-book location 1635
2 'Deutsche Physik', Wikipedia, accessed 28 March 2022, https://en.wikipedia.org/wiki/Deutsche_Physik
3 'Werner Heisenberg – SS Investigation', Wikipedia, accessed 28 March 2022, https://en.wikipedia.org/wiki/Werner_Heisenberg#SS_investigation
4 Mahaffey, *Atomic Awakening*, e-book location 1655
5 Mahaffey, *Atomic Awakening*, e-book location 1998
6 Mariko Oi, 'The Man Who Saved Kyoto from the Atomic Bomb', BBC News, 9 August 2015, https://www.bbc.co.uk/news/world-asia-33755182
7 Mahaffey, *Atomic Awakening*, e-book location 2692
8 Michelle Hall, 'By the Numbers: World War II's atomic bombs', CNN, 6 August 2013, https://edition.cnn.com/2013/08/06/world/asia/btn-atomic-bombs/index.html
9 Mahaffey, *Atomic Awakening*, e-book location 2494
10 Mahaffey, *Atomic Awakening*, e-book location 3439

11 Max Roser, Herre Bastian, and Joe Hasell, 'Nuclear Weapons', Our World in Data, accessed 14 May 2013, https://ourworldindata.org/nuclear-weapons
12 Mahaffey, *Atomic Awakening*, e-book location 3074
13 Mahaffey, *Atomic Awakening*, e-book location 323
14 Mahaffey, *Atomic Awakening*, e-book location 719
15 'World Uranium Mining Production', World Nuclear Association, accessed 11 August 2023, https://www.world-nuclear.org/information-library/nuclear-fuel-cycle/mining-of-uranium/world-uranium-mining-production.aspx
16 'Spent Fuel', Posiva, accessed 16 March 2022, https://www.posiva.fi/en/index/finaldisposal/long-termsafety/spentfuel.html
17 'Nuclear 101: How Does a Nuclear Reactor Work?' Office of Nuclear Energy, 2 August 2023, https://www.energy.gov/ne/articles/nuclear-101-how-does-nuclear-reactor-work
18 'How is Uranium made into Nuclear Fuel?' World Nuclear Association, accessed 28 March 2022, https://www.world-nuclear.org/nuclear-essentials/how-is-uranium-made-into-nuclear-fuel.aspx
19 'Uranium Enrichment', World Nuclear Association, accessed 30 September 2023, https://world-nuclear.org/information-library/nuclear-fuel-cycle/conversion-enrichment-and-fabrication/uranium-enrichment.aspx
20 David Burnham, 'Nuclear Experts Debate "The China Syndrome",' *New York Times*, 18 March 1979, https://www.nytimes.com/1979/03/18/archives/nuclear-experts-debate-the-china-syndrome-but-does-it-satisfy-the.html
21 'Three Mile Island Accident', US Nuclear Regulatory Commission, accessed 9 March 2022, https://www.nrc.gov/docs/ML0825/ML082560250.pdf
22 'Forsmark: how Sweden alerted the world about the danger of the Chernobyl disaster', European Parliament, 15 May 2014, https://www.europarl.europa.eu/news/en/headlines/society/20140514STO47018/forsmark-how-sweden-alerted-the-world-about-the-danger-of-chernobyl-disaster
23 Richard Gray, 'The true toll of the Chernobyl disaster', BBC Future, 26 July 2019, https://www.bbc.com/future/article/20190725-will-we-ever-know-chernobyls-true-death-toll
24 Mahaffey, *Atomic Awakening*, e-book location 4739
25 Bennett, Leonard, and Robert Skjoeldebrand, 'Special Reports: Nuclear Power Development Worldwide Nuclear Power Status and Trends Nuclear's Contribution to Electricity Supply Is Growing,' International Atomic Energy Agency, accessed 22 April 2023, https://www.iaea.org/sites/default/files/publications/magazines/bulletin/bull28-3/28304784045.pdf
26 'The World Nuclear Industry Status Report 2010–2011 (HTML) – Nuclear Power in a Post-Fukushima World 25 Years after the Chernobyl Accident',

World Nuclear Industry Status Report, accessed 22 April 2023, https://www.worldnuclearreport.org/The-World-Nuclear-Industry-Status-51.html

27 BBC News, 'Fukushima: The nuclear disaster that shook the world', 13 March 2021, https://www.youtube.com/watch?v=mUBxtTEOiPI

28 Hannah Ritchie, 'What Are the Safest and Cleanest Sources of Energy?' Our World in Data, 10 February 2020, https://ourworldindata.org/safest-sources-of-energy

29 'Plans for New Nuclear Reactors Worldwide', World Nuclear Association, accessed 6 February 2023, https://world-nuclear.org/information-library/current-and-future-generation/plans-for-new-reactors-worldwide.aspx

30 'International Nuclear Waste Disposal Concepts', World Nuclear Association, accessed 30 September 2023, https://world-nuclear.org/information-library/nuclear-fuel-cycle/nuclear-wastes/international-nuclear-waste-disposal-concepts.aspx

31 Author interview with Pasi Tuohimaa, communications manager at Posiva, 8 February 2022

32 'Final Disposal – Spent Fuel', Posiva, accessed 28 March 2022, https://www.posiva.fi/en/index/finaldisposal/long-termsafety/spentfuel.html

33 Sonal Patel, 'Sweden's Government Approves Construction of Spent Nuclear Fuel Repository', *POWER Magazine*, 1 March 2022, https://www.powermag.com/swedens-government-approves-construction-of-spent-nuclear-fuel-repository/

34 'Policy Paper – Advanced Nuclear Technologies', Department for Business, Energy and Industrial Strategy, accessed 16 March 2022, https://www.gov.uk/government/publications/advanced-nuclear-technologies/advanced-nuclear-technologies

35 Mark Shwartz, 'Stanford-led research finds small modular reactors will exacerbate challenges of highly radioactive nuclear waste', *Stanford News*, 30 May 2022, https://news.stanford.edu/2022/05/30/small-modular-reactors-produce-high-levels-nuclear-waste/

36 Mahaffey, *Atomic Awakening*, e-book location 288

37 Author interview with Dr Melanie Windridge, physicist and founder of Fusion Energy Insights, 31 March 2022

38 Sehila Gonzalez de Vincente *et al.*, 'Overview on the management of radioactive waste from fusion facilities: ITER, demonstration machines and power plants', *Nuclear Fusion* 62, 085001 (March 2022), https://doi.org/10.1088/1741-4326/ac62f7

39 'NIF achieves energy gain', Fusion Energy Insights, 13 December 2022, https://fusionenergyinsights.com/blog/post/nif-achieves-energy-gain

40 Zihan Lin, 'First steady-state, high performance tokamak plasma achieved on EAST in China', Fusion Energy Insights, 27 April 2023, https://fusion-energyinsights.com/blog/post/first-steady-state-high-performance-tokamak-plasma-achieved-on-east-in-china

Chapter 3

1 Brian Hurley, 'How Much Wind Energy Is There? Wind Site Evaluation', Claverton Energy Research Group, 25 March 2009, https://claverton-energy.com/how-much-wind-energy-is-there-brian-hurley-wind-site-evaluation-ltd.html
2 Based on 180,000 TW of solar energy reaching the Earth, which is 180,000 TWh in one hour, exceeding the c. 170,000 TWh consumed globally in 2022 (as per Energy Institute's Statistical Review of World Energy, 2023, 604 Exajoules consumed in 2022)
3 Perlin, *Let it Shine: The 6,000 Year Story of Solar Energy* (Novato: New World Library: 2013), 3–5
4 Perlin, *Let it Shine: The 6,000 Year Story of Solar Energy*, 27–33
5 Paul Collins, 'The Beautiful Possibility', *Cabinet Magazine*, Spring 2002, https://www.cabinetmagazine.org/issues/6/collins.php
6 'Spain CSP Project Development', Solar Paces, accessed 25 April 2022, https://www.solarpaces.org/csp-technologies/csp-potential-solar-thermal-energy-by-member-nation/spain/
7 'Solar Energy: Mapping the Road Ahead', IEA, Paris, 2019, https://www.iea.org/reports/solar-energy-mapping-the-road-ahead
8 Thoubboron, Kerry, 2018. '60 Cell vs. 72 Cell Solar Panels: Which Is Right for You? | EnergySage', Solar News, 30 August 2018, https://news.energysage.com/60-vs-72-cell-solar-panels-which-is-right/
9 Perlin, *Let it Shine: The 6,000 Year Story of Solar Energy*, 303
10 Perlin, *Let it Shine: The 6,000 Year Story of Solar Energy*, 305
11 Perlin, *Let it Shine: The 6,000 Year Story of Solar Energy*, 306
12 'Edmond Becquerel', Physics Today, accessed 17 April 2022, https://physicstoday.scitation.org/do/10.1063/pt.5.031182/full/
13 Perlin, *Let it Shine: The 6,000 Year Story of Solar Energy*, 307–308
14 Perlin, *Let it Shine: The 6,000 Year Story of Solar Energy*, 311
15 Perlin, *Let it Shine: The 6,000 Year Story of Solar Energy*, 310–317
16 'Vanguard Radio Fails to Report; Chemical Battery Believed Exhausted in Satellite – Solar Unit Functioning', *New York Times*, 6 April 1958, https://www.nytimes.com/1958/04/06/archives/vanguard-radio-fails-to-report-chemical-battery-believed-exhausted.html

17 Perlin, *Let it Shine: The 6,000 Year Story of Solar Energy*, 324
18 Perlin, *Let it Shine: The 6,000 Year Story of Solar Energy*, 327–328
19 Perlin, *Let it Shine: The 6,000 Year Story of Solar Energy*, 331–337
20 Perlin, *Let it Shine: The 6,000 Year Story of Solar Energy*, 341
21 T Erge, V.U Hoffmann, and K Kiefer, 2001, 'The German Experience with Grid-Connected PV-Systems', *Solar Energy* 70 (6): 479–87, https://doi.org/10.1016/s0038-092x(00)00143-2
22 5MW for 2,000 homes is 2.5kW per home; a four-person household's electricity needs are roughly 4kW (for UK – this would vary depending on location and other factors)
23 Perlin, *Let it Shine: The 6,000 Year Story of Solar Energy*, 419
24 Goodall, *The Switch: How solar, storage and new tech means cheap power for all* (London: Profile Books, 2016), e-book location 42
25 Perlin, *Let it Shine: The 6,000 Year Story of Solar Energy*, 432
26 'Majority of New Renewables Undercut Cheapest Fossil Fuel on Cost', *International Renewable Energy Agency*, 22 June 2021, https://www.irena.org/newsroom/pressreleases/2021/Jun/Majority-of-New-Renewables-Undercut-Cheapest-Fossil-Fuel-on-Cost
27 'With 2,245 MW of Commissioned Solar Projects, World's Largest Solar Park Is Now at Bhadla', *Mercomindia*, 19 March 2020, https://www.mercomindia.com/world-largest-solar-park-bhadla
28 Author interview with Rahul Gupta, director and founder of Rays Power Experts Pvt Ltd., 15 April 2022
29 'Bhadla Solar Parks Transform India's Energy Landscape', Climate Investment Fund, accessed 21 November 2023, https://www.cif.org/CIF10/india/bhadla
30 Author interview with Amit Singla, vice president India at Ecoppia, 22 April 2022
31 'Solar Energy Current Status', Government of India Ministry of New and Renewable Energy, accessed 11 August 2023, https://mnre.gov.in/solar/
32 Perlin, *Let it Shine: The 6,000 Year Story of Solar Energy*, 408
33 'Fact Check Q&A: is Australia the world leader in household solar power?' The Conversation, 27 March 2016, https://theconversation.com/factcheck-qanda-is-australia-the-world-leader-in-household-solar-power-56670
34 Author interview with Greg Morrison, 30 April 2022
35 'Solar Rebate NSW: List of Government Solar Rebates – 2022', Selectra, accessed 30 April 2022, https://selectra.com.au/energy/guides/rebate/solar-rebate-nsw
36 Charles Q Choi, 'Perovskite-Silicon Pairing Sails Past 30 Percent', 12 July 2023, *IEEE Spectrum*, https://spectrum.ieee.org/tandem-solar-cell-perovskite-silicon

ENDNOTES

Chapter 4

1. Jens Vestergaard, Lotte Brandstrup, Robert Goddard, 'A Brief History of the Wind Turbine Industries in Denmark and the United States', *Academy of International Business (Southeast USA Chapter) Conference Proceedings* (November 2004): 322–327, https://pure.au.dk/ws/files/12741/Windmill_paper1
2. Daniel Yergin, *The Prize: The Epic Quest for Oil, Money, and Power* (New York: Simon & Schuster, 1990), 545
3. 'Why the Windmill was Built', Tvindkraft, accessed 17 March 2022, https://www.tvindkraft.dk/en/history/why-the-windmill-was-built
4. Author visit to Tvindkraft and interviews with Allan Jensen and Britta Jensen, 14 March 2022
5. 'IRENA-GWEC: 30 years of policies for wind energy, Denmark Market Overview', IRENA, accessed 11 August 2023, https://www.irena.org/-/media/Files/IRENA/Agency/Publication/2013/GWEC/GWEC_Denmark.pdf?la=en&hash=C14BEEC4FFEEBA20B2B1928582AA23931F092F48
6. 'Global Energy Review 2021', IEA, accessed 15 March 2022, https://www.iea.org/reports/global-energy-review-2021
7. Xiaoyang Li, 'China's Carbon Neutral Goal', Wood Mackenzie, 30 September 2021, https://www.woodmac.com/news/opinion/wind-power-to-play-a-key-role-in-achieving-chinas-carbon-neutral-goal/
8. Hydro plant produced 12.5kW, assuming LED light bulb consumption of 10W
9. 'History', National Hydropower Association, accessed 21 March 2022, https://www.hydro.org/about/history/
10. 'Statistical Review of World Energy, 2023', Energy Institute, accessed 26 August 2023, https://www.energyinst.org/statistical-review
11. Christina Nunez, 'Hydropower, explained', *National Geographic*, 13 May 2019, https://www.nationalgeographic.com/environment/article/hydropower
12. 'Country profile: China', International Hydropower Association, accessed 18 March 2022, https://www.hydropower.org/country-profiles/china
13. 'Country profile: Norway', International Hydropower Association, accessed 18 March 2022, https://www.hydropower.org/country-profiles/norway
14. Peter Oxley, director *Big, Bigger, Biggest: Dam,* National Geographic, 2009. https://www.dailymotion.com/video/x7nlqzc
15. Philip Ball, *The Water Kingdom: A Secret History of China* (London: The Bodley Head, 2016), 465
16. L. Patricia Kite, *Building the Three Gorges Dam* (Oxford: Raintree, 2011), 9
17. L. Patricia Kite, *Building the Three Gorges Dam,* 20

18 *Big, Bigger, Biggest: Dam*, National Geographic, 2009
19 Marza Hvistendahl, 'China's Three Gorges Dam: An Environmental Catastrophe?' *Scientific American*, 25 March 2008, https://www.scientificamerican.com/article/chinas-three-gorges-dam-disaster/
20 L. Patricia Kite, *Building the Three Gorges Dam*, 23
21 L. Patricia Kite, *Building the Three Gorges Dam*, 46–49
22 Christina Nunez, 'Hydropower, explained', *National Geographic*, 13 May 2019, https://www.nationalgeographic.com/environment/article/hydropower (accessed 21 March 2022)
23 'Hydroelectric dams emit a billion tonnes of greenhouse gases a year, study finds', *Guardian*, 14 November 2016, https://www.theguardian.com/global-development/2016/nov/14/hydroelectric-dams-emit-billion-tonnes-greenhouse-gas-methane-study-climate-change
24 'History and Controversy of the Three Gorges Dam', Britannica, accessed 21 March 2022, https://www.britannica.com/topic/Three-Gorges-Dam/History-and-controversy-of-the-Three-Gorges-Dam
25 'Hydropower Special Market Report', IEA, accessed 28 April 2023, https://www.iea.org/reports/hydropower-special-market-report
26 Chris Martin, 'Wind Turbine Blades Can't Be Recycled, So They're Piling Up in Landfills', *Bloomberg*, 5 February 2020, https://www.bloomberg.com/news/features/2020-02-05/wind-turbine-blades-can-t-be-recycled-so-they-re-piling-up-in-landfills
27 Heidi Vella, 'An Industry in the Making: Diverting Wind Turbine Blades from Landfill', *E&T Magazine*, 13 October 2022, https://eandt.theiet.org/content/articles/2022/10/an-industry-in-the-making-diverting-wind-turbine-blades-from-landfill/
28 'Siemens Gamesa pioneers wind circularity: launch of world's first recyclable wind turbine blade for commercial use offshore', Siemens Gamesa, 7 September 2021, https://www.siemensgamesa.com/en-int/newsroom/2021/09/launch-world-first-recyclable-wind-turbine-blade

Chapter 5

1 'Energy imports, net (% of energy use) – Country Ranking', Index Mundi, accessed 6 February 2023, https://www.indexmundi.com/facts/indicators/EG.IMP.CONS.ZS/rankings
2 Peter Thorsheim, 'The Paradox of Smokeless Fuels: Gas, Coke and the Environment in Britain, 1813–1949', *Environment and History* 8, no. 4 (November 2002): 381–401

ENDNOTES

3 'W. Wood Prince, 83, of Chicago; Businessman Adopted by Cousin', *New York Times*, 30 January 1998, https://www.nytimes.com/1998/01/30/business/w-wood-prince-83-of-chicago-businessman-adopted-by-cousin.html
4 Peter G. Noble, 'A Short History of LNG Shipping 1959–2009', Texas Section – Society of Naval Architects and Marine Engineers, 10 February 2009, https://higherlogicdownload.s3.amazonaws.com/SNAME/1dcdb863-8881-4263-af8d-530101f64412/UploadedFiles/c3352777fcaa4c4daa8f-125c0a7c03e9.pdf
5 Gordon Shearer and Michael D. Tusiani, *LNG Fuel for a Changing World: A non-technical guide* (Tulsa, Oklahoma: PennWell Corporation, 2016)
6 'Number of liquefied natural gas storage vessels worldwide from 2010 to 2021', Statista, accessed 13 November 2022, https://www.statista.com/statistics/468412/global-lng-tanker-fleet/
7 Author interview with Captain Tatjana Pletena, BW Everett, 31 October 2021
8 'The Rise of Female Cruiseship Captains', *Atlantic Pacific*, 17 December 2019, https://atlanticpacific.co/the-rise-of-female-cruise-ship-captains/
9 Based on LNG density of 450 kg/m^3, calorific value of 55MJ/kg, and average energy bill for a UK three-bedroom home consuming 12MWh per year, https://www.britishgas.co.uk/energy/guides/average-bill.html
10 'My Average Gas Consumption – Calculate & Estimate', Wolf, accessed 14 May 2023, https://www.wolf.eu/en-de/advisor/average-gas-consumption (12–15MWh per year for 100m2, 1MWh = 91 standard cubic metres of natural gas)
11 'Statistical Review of World Energy, 2023', Energy Institute, accessed 26 August 2023, https://www.energyinst.org/statistical-review
12 'Montreal, Maine & Atlantic Railway (MMA) Derailment in Lac-Mégantic, Quebec, press release', Montreal Maine and Atlantic Railway, 6 July 2013, https://www.canoils.com/Universal/View.aspx?type=Story&id=30221
13 'Lac-Mégantic runaway train and derailment investigation summary', Transportation Safety Board of Canada, accessed 6 February 2023, https://www.tsb.gc.ca/eng/rapports-reports/rail/2013/r13d0054/r13d0054-r-es.html
14 Lucy Easthope, *When the Dust Settles* (London: Hodder & Stoughton, 2022), 154
15 Rhodes, *Energy: A Human History*, 65
16 Rhodes, *Energy: A Human History*, 64
17 'Chronology of America's Freight Railroads', Association of American Railroads, accessed 6 September 2023, https://www.aar.org/chronology-of-americas-freight-railroads/
18 'Tank cars', Alaska Rails, accessed 6 February 2023, https://www.alaskarails.org/fp/tankcars.html

19 'Refinery Receipts of Crude Oil by Method of Transportation', US Energy Information Administration, accessed 6 February 2023, https://www.eia.gov/dnav/pet/pet_pnp_caprec_dcu_nus_a.htm
20 Georgia Simcox, 'Fuel tanker explodes into a huge fireball after crashing on Michigan highway in terrifying dashboard camera video', *Daily Mail*, 14 July 2021, https://www.dailymail.co.uk/news/article-9786815/Fuel-tanker-explodes-huge-fireball-crashing-Michigan-highway.html
21 'Band leads first petrol tanker in Hamilton', Hamilton Libraries, accessed 6 February 2023, https://heritage.hamiltonlibraries.co.nz/objects/50/band-leads-first-petrol-tanker-in-hamilton

Chapter 6

1 Phil Hopkins, 'Pipelines: Past, Present, and Future', Penspen Integrity, 8 March 2007, https://www.penspen.com/wp-content/uploads/2014/09/past-present-future.pdf
2 Hopkins, 'Pipelines: Past, Present, and Future'
3 Madelon L. Finkel, *Pipeline Politics: Assessing the Benefits and Harms of Energy Policy* (Santa Barbara, California: Praeger, 2018)
4 Madelon L. Finkel, 'Pipeline Politics book talk', Cornell University, 29 November 2018, https://www.cornell.edu/video/pipeline-politics-energy-policy-book-talk-by-madelon-l-finkel
5 Around the equatorial circumference of the Earth, 40,000 kilometres
6 Patrick Wintour, 'Nord Stream 2: how Putin's pipeline paralysed the West', *Guardian*, 23 December 2021, https://www.theguardian.com/world/2021/dec/23/nord-stream-2-how-putins-pipeline-paralysed-the-west
7 'Secure Energy for Europe: the Nord Stream Pipeline project', Nord Stream AG, May 2013, https://www.nord-stream.com/media/documents/pdf/en/2014/04/secure-energy-for-europe-full-version.pdf, 22
8 'World's Longest Subsea Pipeline – Megastructures', 9 October 2020, YouTube https://www.youtube.com/watch?v=iGBPgtomdjA (Note: this video is about the Langeled pipeline, but the pipe-laying method was similar for Nord Stream)
9 'Secure Energy for Europe: the Nord Stream Pipeline project', Nord Stream AG, May 2013, 177
10 Thomas Forde, 'Diving 180 Metres under the Surface', Stavanger Aftenblad, 23 August 2012, https://www.aftenbladet.no/aenergi/i/zkA3K/diving-180-metres-under-the-surface
11 'Secure Energy for Europe: the Nord Stream Pipeline project', Nord Stream AG, May 2013 (p. 177)

ENDNOTES

12 'Secure Energy for Europe: the Nord Stream Pipeline project', Nord Stream AG, May 2013 (p. 177)
13 'Secure Energy for Europe: the Nord Stream Pipeline project', Nord Stream AG, May 2013 (p. 177)
14 'Secure Energy for Europe: the Nord Stream Pipeline project', Nord Stream AG, May 2013 (p. 121)
15 'Nord Stream Sabotage One Year On: What to Know about the Attack', Al Jazeera, accessed 14 October 2023, https://www.aljazeera.com/news/2023/9/23/what-we-know-about-the-nord-stream-sabotage-one-year-on
16 Nerijus Adomaitis, 'Q+A: What Is Known about the Nord Stream Gas Pipeline Explosions', Reuters, 26 September 2023, https://www.reuters.com/world/europe/qa-what-is-known-about-nord-stream-gas-pipeline-explosions-2023-09-26/
17 'How Europe Can Cut Natural Gas Imports from Russia Significantly within a Year – News', *IEA*, 3 March 2022, https://www.iea.org/news/how-europe-can-cut-natural-gas-imports-from-russia-significantly-within-a-year
18 Wintour, 'Nord Stream 2: how Putin's pipeline paralysed the West'
19 Jeff Mason, 'Trump lashes Germany over gas pipeline deal, calls it Russia's "captive"', Reuters, 11 July 2018, https://www.reuters.com/article/us-nato-summit-pipeline-idUSKBN1K10VI
20 Wintour, 'Nord Stream 2: how Putin's pipeline paralysed the West'
21 Madelon L. Finkel, *Pipeline Politics: Assessing the Benefits and Harms of Energy Policy*
22 Robert F. Kennedy Jr., 'Why the Arabs don't want us in Syria', *Politico Magazine*, 22 February 2016, https://www.politico.com/magazine/story/2016/02/rfk-jr-why-arabs-dont-trust-america-213601/
23 'Oil Pipelines', Britannica, accessed 19 February 2022, https://www.britannica.com/place/Iraq/Oil-pipelines
24 Madelon L Finkel, *Pipeline Politics: Assessing the Benefits and Harms of Energy Policy*
25 Andy Balaskovitz, 'Biden Cancels Keystone XL Pipeline on First Day in Office', *Energy News Network*, 21 January 2021, https://energynews.us/digests/biden-cancels-keystone-xl-pipeline-on-first-day-in-office/
26 Holly Honderich, 'Keystone XL: Why I fought for – or against – the pipeline', BBC News, 27 January 2021, https://www.bbc.co.uk/news/world-us-canada-55816229
27 Jonathan L. Ramseur, Richard K. Lattanzio, Linda Luther, Paul W. Parfomak and Nicole T. Carter, 'Oil Sands and the Keystone XL Pipeline: Background and Selected Environmental Issues', Congressional Research Services, 14 April 2014, https://sgp.fas.org/crs/misc/R42611.pdf

28 Rebecca Rooney et al., 'Oil Sands Mining and Reclamation Cause Massive Loss of Peatland and Stored Carbon', *Proceedings of the National Academy of Sciences* 109, 13 (March 2012): 4933–4937. https://www.pnas.org/doi/10.1073/pnas.1117693108

29 Nina Lakhani, '"No More Broken Treaties": Indigenous Leaders Urge Biden to Shut down Dakota Access Pipeline', *Guardian*, 21 January 2021, https://www.theguardian.com/us-news/2021/jan/21/dakota-access-pipeline-joe-biden-indigenous-environment

30 Honderich, 'Keystone XL: Why I fought for – or against – the pipeline'

31 'Alberta Mining and Oil and Gas Extraction Industry Profile 2020', Labour and Immigration – Government of Alberta, 14 September 2021, https://open.alberta.ca/dataset/f4f39b9e-48cb-4f6a-b491-25ee6f9c281e/resource/8db15e6c-5826-4ac5-b804-675e95867e9e/download/lbr-alberta-mining-and-oil-and-gas-extraction-industry-profile-2020.pdf

32 Jillian Ambrose, 'Prospectors Hit the Gas in the Hunt for "White Hydrogen"', *Guardian*, 12 August 2023, https://www.theguardian.com/environment/2023/aug/12/prospectors-hit-the-gas-in-the-hunt-for-white-hydrogen

33 'Geopolitics of the Energy Transformation: The Hydrogen Factor', *International Renewable Energy Agency*, 15 January 2022, https://www.irena.org/publications/2022/Jan/Geopolitics-of-the-Energy-Transformation-Hydrogen

34 Author interview with Sarah Kimpton, 23 December 2021

Chapter 7

1 Mike Winchell, *The Electric War: Edison, Tesla, Westinghouse, and the Race to Light the World* (Henry Holt & Co, 2019)

2 Mike Winchell, *The Electric War: Edison, Tesla, Westinghouse, and the Race to Light the World*

3 Marco Alverà, *The Hydrogen Revolution: A blueprint for the future of clean energy* (London: Hodder & Stoughton, 2021), 53

4 'Powershop Australia Demand Response Program', Australian Renewable Energy Agency, accessed 8 February 2023, https://arena.gov.au/knowledge-bank/may-2020-powershop-demand-response-program/

5 'Applying Behavioural Insights to Powershop's Curb Your Power program', Australian Renewable Energy Agency, accessed 8 February 2023, https://arena.gov.au/knowledge-bank/applying-behavioural-insights-to-powershops-curb-your-power-program/

6 John Ydstie, 'A Country Divided: Japan's Electric Bottleneck', NPR, 24 March 2011, https://text.npr.org/134828205
7 'Japan's Power Grid: Interconnections', Shulman Advisory, accessed 8 February 2023, https://shulman-advisory.com/2020/06/05/japans-power-grid-interconnections/
8 Gretchen Bakke, *The Grid: The Fraying Wires Between Americans and Our Energy Future* (Bloomsbury, 2016), Chapter 2: How the Grid Got its Wires
9 'News release: New east-west HVDC commissioned, interconnecting Nagano and Gifu prefectures', Toshiba, 1 April 2021: https://www.toshiba-energy.com/en/info/info2021_0401.htm
10 'Gotland, Sweden', NKT, accessed 8 February 2023, https://www.nkt.com/references/gotland-sweden
11 Molly Lempriere, 'China's mega transmission lines', *Power Technology*, 6 March 2019, https://www.power-technology.com/features/chinas-mega-transmission-lines/
12 Marc Champion, 'The Future of Power Is Transcontinental, Submarine Supergrids', *Bloomberg*, 9 June 2021, https://www.bloomberg.com/news/features/2021-06-09/future-of-world-energy-lies-in-uhvdc-transmission-lines
13 'Mongolia's Vast Renewable Energy Resources Can Power Sustainable Development', International Renewable Energy Agency, 21 March 2016, https://www.irena.org/newsroom/pressreleases/2016/Mar/Mongolias-Vast-Renewable-Energy-Resources-Can-Power-Sustainable-Development
14 'Electricity, explained', U.S. Energy Information Administration, accessed 8 February 2023, https://www.eia.gov/energyexplained/electricity/electricity-in-the-us-generation-capacity-and-sales.php
15 'Lifetime Achievement: Tracy Greenwood, El Dorado Hills, California', Flickr, accessed 8 February 2023, https://www.flickr.com/photos/mypubliclands/48050840396
16 Andrew Kuhn, 'Briceburg Fire fully contained at 5,563 acres, according to fire officials', *Merced Sun-Star*, 28 October 2019, https://www.mercedsunstar.com/latest-news/article236749408.html
17 Elisa Wood, 'What is a microgrid?' Microgrid Knowledge, 28 March 2020, https://microgridknowledge.com/microgrid-defined/
18 Author interview with Michele Nesbit, co-founder and chief operating officer of BoxPower, 20 July 2021
19 '2021 Wildfire Mitigation Plan report', Pacific Gas and Electric Company (PG&E), 5 February 2021, https://www.pge.com/pge_global/common/pdfs/safety/emergency-preparedness/natural-disaster/wildfires/wildfire-mitigation-plan/2021-Wildfire-Safety-Plan.pdf
20 PG&E 2021 Wildfire Mitigation Plan states that 10% of customers live in HFTD areas; company's website states that they have 5.5 million electricity customers.

21 'Puerto Rico issues new data on Hurricane Maria deaths', NBC News, 13 June 2018, https://www.nbcnews.com/health/health-news/puerto-rico-issues-new-data-hurricane-maria-deaths-n882816

Chapter 8

1. Author interview with Cara MacEachern and Trevor Schulz, 28 October 2021
2. The Engineering Mindset, 'How A Car Battery Works – Basic Working Principle', 24 August 2020, https://www.youtube.com/watch?v=VnPRX5zQWLw.
3. 'Lithium-ion Batteries: Timeline', Naval Technology, 15 June 2020, https://www.naval-technology.com/comment/lithium-ion-lib-timeline/
4. Mogalahalli V. Reddy et al., 'Brief History of Early Lithium-Battery Development', *Materials* 13, no. 8 (17 April 2020): 1884, https://doi.org/10.3390/ma13081884
5. 'Recycling used lead-acid batteries: health considerations', World Health Organization, 11 January 2017, https://www.who.int/publications/i/item/9789241512855
6. Tony Barboza, 'Story So Far: How a Battery Recycler Contaminated L.A. – Area Homes for Decades', *Los Angeles Times*, 21 December 2015, https://www.latimes.com/local/lanow/la-me-exide-cleanup-story-so-far-20151121-story.html
7. 'Electric Vehicles', IEA, accessed 8 February 2023, https://www.iea.org/reports/electric-vehicles
8. 'Developing Countries Pay Environmental Cost of Electric Car Batteries', UNCTAD, 22 July 2020, https://unctad.org/news/developing-countries-pay-environmental-cost-electric-car-batteries
9. Amit Katwala, 'The Spiralling Environmental Cost of Our Lithium Battery Addiction', *WIRED UK*, 5 August 2018, https://www.wired.co.uk/article/lithium-batteries-environment-impact
10. 'Lithium Statistics and Information – Mineral Commodity Summaries', *U.S. Geological Survey*, January 2021, https://pubs.usgs.gov/periodicals/mcs2021/mcs2021-lithium.pdf
11. Catherine Early, 'The New "Gold Rush" for Green Lithium', BBC Future, 25 November 2020, https://www.bbc.com/future/article/20201124-how-geothermal-lithium-could-revolutionise-green-energy
12. 'Cobalt Statistics and Information – Mineral Commodity Summaries', *U.S. Geological Survey*, January 2021, https://pubs.usgs.gov/periodicals/mcs2021/mcs2021-cobalt.pdf
13. Katharine Sanderson, 'The Long Road to Sustainable Lithium-Ion Batteries', Chemistry World, July 5, 2021, https://www.chemistryworld.

com/features/the-long-road-to-sustainable-lithium-ion-batteries/4013838. article
14 'Green Deal: Sustainable batteries for a circular and climate neutral economy', European Commission, 10 December 2020, https://ec.europa.eu/commission/presscorner/detail/en/ip_20_2312
15 'SPR Storage Sites', U.S. Department of Energy, accessed 8 February 2023, https://www.energy.gov/fecm/strategic-petroleum-reserve-4
16 Ad van Wijk, Frank Wouters, 'Hydrogen – The Bridge Between Africa and Europe', *Shaping an Inclusive Energy Transition*, (June 2021): 91–119. https://doi.org/10.1007/978-3-030-74586-8_5
17 'Proving the viability of underground hydrogen storage', *IEA*, 22 October 2021, https://www.iea.org/articles/proving-the-viability-of-underground-hydrogen-storage
18 Liana Paraschaki, 'Beira, the Cailleach, Queen of Winter', Folklore Scotland, accessed 6 September 2023, https://folklorescotland.com/beira-the-cailleach/
19 'History', Visit Cruachan, accessed 8 February 2023, https://www.visitcruachan.co.uk/history/
20 Author visit to Cruachan power station and interviews with Ian Kinnaird, Scottish Assets and Generation Engineering Director, and Sarah Cameron, Community Manager (Scotland) for Drax Group, 4 August 2021
21 'Cruachan Power Station', Drax, accessed 8 February 2023, https://www.drax.com/about-us/our-sites-and-businesses/cruachan-power-station/
22 'Scottish Munros and Munro Bagging', Visit Scotland, accessed 8 February 2023, https://www.visitscotland.com/see-do/active/walking/munro-bagging/
23 'Drax Submits Application to Expand Iconic "Hollow Mountain" Power Station', *Drax*, 17 May 2022, https://www.drax.com/press_release/drax-submits-application-to-expand-iconic-hollow-mountain-power-station/
24 Priyanka Shrestha, 'Australia's First Pumped Hydro Project in 37 Years Reaches Financial Close', *Energy Live News*, 27 May 2021, https://www.energylivenews.com/2021/05/27/australias-first-pumped-hydro-project-in-37-years-reaches-financial-close/
25 '250MW Kidston Pumped Storage Hydro Project,' Genex Power, accessed 8 February 2023, https://genexpower.com.au/250mw-kidston-pumped-storage-hydro-project/
26 Max Hall, 'Israel Prepares 800 MW of Pumped Hydro Storage', Pv Magazine International, 24 June 2020, https://www.pv-magazine.com/2020/06/24/israel-prepares-800-mw-of-pumped-hydro-storage/

Chapter 9

1. As of 2023, the Museum of London is in the process of relocating to Smithfield Market, London
2. 'Energy Consumption in Households', Eurostat – Statistics Explained, accessed 8 February 2023, https://ec.europa.eu/eurostat/statistics-explained/index.php?title=Energy_consumption_in_households
3. 'Heating – process or system', Britannica, accessed 8 February 2023, https://www.britannica.com/technology/heating-process-or-system
4. 'District Heating and Cooling', Macmillan Encyclopedia of Energy, accessed 12 February 2023, https://www.encyclopedia.com/environment/encyclopedias-almanacs-transcripts-and-maps/district-heating-and-cooling
5. 'Gumry Hotel – 1895', Denver Fire Journal, accessed 12 February 2023, http://denverfirejournal.blogspot.com/2014/05/gumry-hotel-disaster.html
6. 'Gumry Hotel Explosion', Colorado Encyclopedia, accessed 12 February 2023, https://coloradoencyclopedia.org/article/gumry-hotel-explosion
7. 'Gumry Hotel – 1895', *Denver Fire Journal*, accessed 12 February 2023, http://denverfirejournal.blogspot.com/2014/05/gumry-hotel-disaster.html
8. 'The Future of Cooling: Opportunities for Energy Efficient Air Conditioning', IEA, accessed 12 February 2023, https://www.iea.org/reports/the-future-of-cooling
9. Jennifer F. Bobb *et al.*, 'Cause-Specific Risk of Hospital Admission Related to Extreme Heat in Older Adults', *JAMA* 312, no. 24 (24 December 2014): 2659, https://doi.org/10.1001/jama.2014.15715
10. 'The Future of Cooling: Opportunities for Energy Efficient Air Conditioning'
11. Author interview with Mahendra Patel, Chairman and Managing Director of the Mamata Group, 19 July 2021
12. Author interview with Robert Edwards, co-founder and chief scientific officer of Solar Polar Limited, 28 June 2021
13. 'Mam[a]ta Energy unveils first solar thermal AC', Business Standard, 7 March 2013, https://www.business-standard.com/article/companies/mamta-energy-unveils-first-solar-thermal-ac-106042001069_1.html
14. 'Food Loss and Waste Database', Food and Agriculture Organisation of the United Nations, accessed 12 February 2023, http://www.fao.org/food-loss-and-food-waste/flw-data
15. Author visit to Hockerton Housing Project and interview with Simon Tilley, director of Hockerton Housing Project, 17 July 2021

Chapter 10

1. 'Turkey – Agriculture', International Trade Administration, 26 July 2022, https://www.trade.gov/country-commercial-guides/turkey-agriculture
2. Thomas Hager, *The Alchemy of Air: A Jewish Genius, a Doomed Tycoon, and the Scientific Discovery That Fed the World but Fueled the Rise of Hitler* (Broadway Books, 2008), e-book location 87
3. Hager, *The Alchemy of Air*, 8
4. Hager, *The Alchemy of Air*, 5
5. Hager, *The Alchemy of Air*, 51
6. Hager, *The Alchemy of Air*, 51
7. Hager, *The Alchemy of Air*, 62
8. Hager, *The Alchemy of Air*, 69–71
9. Hager, *The Alchemy of Air*, 72–73
10. Hager, *The Alchemy of Air*, 74
11. Hager, *The Alchemy of Air*, 67–68
12. Hager, *The Alchemy of Air*, 89–91
13. Hager, *The Alchemy of Air*, 100
14. Hager, *The Alchemy of Air*, 93
15. Hager, *The Alchemy of Air*, 104
16. Hager, *The Alchemy of Air*, 122
17. Hager, *The Alchemy of Air*, 140–141
18. Hager, *The Alchemy of Air*, 151
19. Hager, *The Alchemy of Air*, 144
20. Hager, *The Alchemy of Air*, 168
21. Hager, *The Alchemy of Air*, 157 – 159, 163
22. Hager, *The Alchemy of Air*, 156
23. Hager, *The Alchemy of Air*, 163
24. Hager, *The Alchemy of Air*, 234
25. Hager, *The Alchemy of Air*, 235–236
26. Hager, *The Alchemy of Air*, 241
27. Hager, *The Alchemy of Air*, 246
28. Hager, *The Alchemy of Air*, 279
29. Hager, *The Alchemy of Air*, 268
30. Hager, *The Alchemy of Air*, 271
31. Hager, *The Alchemy of Air*, 273, 276
32. John Vidal, 'Bird Flu "an Urgent Warning to Move Away from Factory Farming"', *Guardian*, 6 October 2022, https://www.theguardian.com/environment/2022/oct/06/bird-flu-an-urgent-warning-to-move-away-from-factory-farming
33. 'Ammonia: Zero Carbon Fertilizer, Fuel and Energy Store', Royal Society,

accessed 12 February 2023, https://royalsociety.org/-/media/policy/projects/green-ammonia/green-ammonia-policy-briefing.pdf

34 'New IEA study examines the future of the ammonia industry amid efforts to reach net zero emissions', *IEA*, 11 October 2021, https://www.iea.org/news/new-iea-study-examines-the-future-of-the-ammonia-industry-amid-efforts-to-reach-net-zero-emissions

35 'Ammonia Technology Roadmap', IEA, accessed 12 February 2023, https://www.iea.org/reports/ammonia-technology-roadmap

36 Colin Gagg, 'Cement and Concrete as an Engineering Material: An Historic Appraisal and Case Study Analysis', *Engineering Failure Analysis* 40 (May 1, 2014): 114–40, https://doi.org/10.1016/j.engfailanal.2014.02.004

37 Johanna Lehne, Felix Preston, 'Making Concrete Change: Innovation in Low Carbon Cement and Concrete', Chatham House, 13 June 2018, https://www.chathamhouse.org/2018/06/making-concrete-change-innovation-low-carbon-cement-and-concrete

38 Robbie M. Andrew, 'Global CO_2 Emissions from Cement Production', *Earth System Science Data* 10, no. 1 (26 January 2018): 195–217, https://doi.org/10.5194/essd-10-195-2018

39 Michael A. Nisbet, Medgar L. Marceau, Martha G. VanGeem, 'Environmental Life Cycle Inventory of Portland Cement Concrete', Portland Cement Association, July 2002, https://web.archive.org/web/20170516142547/https://www.nrmca.org/taskforce/item_2_talkingpoints/sustainability/sustainability/sn2137a.pdf

40 Author interview with Diana Casey, executive director of energy and climate change at the Mineral Products Association, 13 June 2022

41 Hilary Clarke, 'The Pantheon: still the world's largest unreinforced concrete dome', *E&T Magazine*, 15 March 2022, https://eandt.theiet.org/content/articles/2022/03/the-pantheon-still-the-world-s-largest-reinforced-concrete-dome/

42 Lucy Rodgers, 'Climate Change: The Massive CO_2 Emitter You May Not Know About', BBC News, 17 December 2018, https://www.bbc.co.uk/news/science-environment-46455844

43 Author visit to a UK cement site, 25 February 2020

44 Lehne and Preston, 'Making Concrete Change: Innovation in Low Carbon Cement and Concrete'

45 'Global Thermal Energy Intensity of Clinker Production by Fuel in the Net Zero Scenario, 2010-2030 – Charts – Data & Statistics', IEA, accessed 10 August 2023, https://www.iea.org/data-and-statistics/charts/global-thermal-energy-intensity-of-clinker-production-by-fuel-in-the-net-zero-scenario-2010-2030

46 Marcus Fairs, 'Concrete Construction "Offsets around One Half" of Emissions Caused by Cement Industry', Dezeen, 21 August 2021, https://

www.dezeen.com/2021/08/24/concrete-construction-offsets-emissions-cement-industry-ipcc/
47 'New continuous concrete pour record set at Hinkley Point C', Hanson, 28 June 2019, https://www.hanson.co.uk/en/about-us/news-and-events/continuous-concrete-pour-record-set-at-hinkley-point-c
48 'Industry', IEA, accessed 12 February 2023, https://www.iea.org/topics/industry
49 Robert F. Service, 'Can the World Make the Chemicals it Needs Without Oil?' *Science*, 19 September 2019, https://www.science.org/content/article/can-world-make-chemicals-it-needs-without-oil

Chapter 11

1 'World Balance 2020', IEA, accessed 2 October 2022, https://www.iea.org/sankey/#?c=World&s=Balance
2 'Transport', Energy Institute, accessed 2 October 2022, https://www.energyinst.org/exploring-energy/topic/transport
3 Patrick Moriarty and Damon Honnery, 'Global Transport Energy Consumption', in *Alternative Energy and Shale Gas Encyclopedia*, ed. Jay H. Lehr, Jack Keeley and Thomas B. Kingery (Hoboken: John Wiley & Sons, 2016), 651–56, https://doi.org/10.1002/9781119066354.ch61
4 'World Final Consumption 2020', IEA, accessed 2 October 2022, https://www.iea.org/sankey/#?c=World&s=Final%20consumption
5 'Global energy-related CO2 emissions by sector', IEA, accessed 12 February 2023, https://www.iea.org/data-and-statistics/charts/global-energy-related-co2-emissions-by-sector
6 'Bertha Benz', Britannica, accessed 6 November 2023, https://www.britannica.com/biography/Bertha-Benz
7 Elena Luchian, 'Mercedes-Benz Museum – How Driving Licence Used to Look', Mercedes Blog, 15 August 2020, https://mercedesblog.com/mercedes-benz-museum-how-driving-licence-used-to-look/
8 'The World's First Long-Distance Car Journey', BBC Reel, 8 August 2022, https://www.bbc.com/reel/video/pocr5jrj/the-world-s-first-long-distance-car-journey?utm_source=taboola
9 'Bertha Benz Memorial Route', Wikipedia, accessed 12 February 2023, https://en.wikipedia.org/wiki/Bertha_Benz_Memorial_Route
10 Yergin, *The Prize: The Epic Quest for Oil, Money, and Power*, 542
11 'Global EV Outlook 2021', IEA, accessed 23 September 2022, https://www.iea.org/reports/global-ev-outlook-2021

12 'The History of the Electric Car', U.S. Department of Energy, 14 September 2015, https://www.energy.gov/articles/history-electric-car
13 'Global EV Outlook 2023', IEA, accessed 7 September 2023, https://www.iea.org/reports/global-ev-outlook-2023
14 'The New Toyota Mirai', Toyota, accessed 6 September 2023, https://newsroom.toyota.eu/the-new-toyota-mirai/
15 Leigh Collins, 'The Number of Hydrogen Fuel-Cell Vehicles on the World's Roads Grew by 40% in 2022, Says IEA Report', Hydrogen Insight, 2 May 2023, https://www.hydrogeninsight.com/transport/the-number-of-hydrogen-fuel-cell-vehicles-on-the-worlds-roads-grew-by-40-in-2022-says-iea-report/2-1-1444069
16 Johannes Pleschberger, 'Europe Expands Night Train Network as Rail Becomes More Popular', Euronews, 16 December 2020, https://www.euronews.com/2020/12/16/europe-expands-night-train-network-as-rail-becomes-more-popular
17 'Climate Change: Should You Fly, Drive or Take the Train?' BBC News, 24 August 2019, https://www.bbc.co.uk/news/science-environment-49349566
18 Author interview with Robyn Huiting, Industrial Design student at Eindhoven University of Technology, 31 August 2022
19 'Falcon Electric Aviation', Falcon Electric Aviation, accessed 12 February 2023, https://falconea.nl/
20 Brian Lowe, 'Cessna 150 Safety Review', Flying in Ireland, January 2014, https://www.gasci.ie/uploads/1/2/9/1/12917568/flying_in_ireland_c150_jan14.pdf
21 Alastair Bland, 'How Bad Is Air Travel for the Environment?' *Smithsonian Magazine*, 26 September 2012, https://www.smithsonianmag.com/travel/how-bad-is-air-travel-for-the-environment-51166834
22 'ZEROe – towards the world's first hydrogen-powered commercial aircraft', Airbus, accessed 9 September 2023, https://www.airbus.com/en/innovation/low-carbon-aviation/hydrogen/zeroe
23 Donovan Hohn, 'The Great Escape: The Bath Toys That Swam the Pacific', *Guardian*, 20 September 2017, https://www.theguardian.com/environment/2012/feb/12/great-escape-bath-toys-pacific
24 Olivia Yasukawa, 'Ducks Overboard! What Happens to Goods Lost at Sea?' CNN, 9 October 2013, http://edition.cnn.com/2013/10/09/business/goods-lost-at-sea/index.html
25 Marianne Guenot, 'The Company behind the Suez Canal Blockage Spilled 28,800 Plastic Toys into the Ocean in the 1990s', Business Insider, 26 March 2021, https://www.businessinsider.com/suez-canal-blockage-ever-green-marine-once-spilled-plastic-toys-ocean-2021-3?international=true&r=US&IR=T

ENDNOTES

26 'Green corridors: A lane for zero-carbon shipping', McKinsey, 21 December 2021, https://www.mckinsey.com/business-functions/sustainability/our-insights/green-corridors-a-lane-for-zero-carbon-shipping
27 Dominik Englert and Andrew Losos, 'Zero-Emission Shipping: What's in It for Developing Countries?' World Bank Blogs, 24 February 2020, https://blogs.worldbank.org/transport/zero-emission-shipping-whats-it-developing-countries
28 Joe Hernandez, 'A Deal That Lets Ukraine Export Grain during Its War with Russia Is about to Expire', Georgia Public Broadcasting, 16 July 2023, https://www.gpb.org/news/2023/07/16/deal-lets-ukraine-exportgrain-during-its-war-russia-about-expire
29 Author interview with Professor Michael Vahs, professor for ship operation and simulation at Hochschule Emden/Leer, University of Applied Sciences, 13 September 2022
30 'Marshall Islands: New Climate Study Visualizes Confronting Risk of Projected Sea Level Rise', World Bank, 29 October 2021, https://www.worldbank.org/en/news/press-release/2021/10/29/marshall-islands-new-climate-study-visualizes-confronting-risk-of-projected-sea-level-rise

Epilogue

1 'Number of internet and social media users as of April 2023', Statista, 29 August 2023, https://www.statista.com/statistics/617136/digital-population-worldwide/
2 Aimee Ross and Lorna Christie, 'Research briefing: Energy consumptions of ICT', UK Parliament, 1 September 2022, https://post.parliament.uk/research-briefings/post-pn-0677/
3 'What Is AI?' McKinsey & Company, 24 April 2023, https://www.mckinsey.com/featured-insights/mckinsey-explainers/what-is-ai
4 Sarah LaBrecque, 'Logistics firms plot a sea-change in sustainability through AI and automation,' *Reuters*, 4 January 2024, https://www.reuters.com/sustainability/climate-energy/logistics-firms-plot-sea-change-sustainability-through-ai-automation-2024-01-04/
5 Carl Elkin, Sims Witherspoon, 'Machine Learning Can Boost the Value of Wind Energy', Google DeepMind, 26 February 2019, https://www.deepmind.com/blog/machine-learning-can-boost-the-value-of-wind-energy
6 'ESO and the Alan Turing Institute Use Machine Learning to Help Balance the GB Electricity Grid', National Grid ESO, 25 July 2019, https://

www.nationalgrideso.com/news/eso-and-alan-turing-institute-use-machine-learning-help-balance-gb-electricity-grid

7 Danae Kyriakopoulou, 'What Opportunities and Risks Does AI Present for Climate Action?' London School of Economics and Grantham Research Institute on Climate Change and the Environment, 4 July 2023, https://www.lse.ac.uk/granthaminstitute/explainers/what-opportunities-and-risks-does-ai-present-for-climate-action/

Index

air conditioning 222–3
air travel 269–74
al-Kasim, Farouk 40
Algeria 117–19
Arab–Israeli war (1973) 92–3
Argentina 193–4
artificial intelligence (AI) 291–2
Artsimovich, Lev 67
Asahi Kasei 192
Assad, Bashar al- 149–50
Australia
    lithium in 194
    solar energy in 80–1, 88–90
    and 'Curb Your Power'
        programme 172–4

batteries
    end-of-life options 195–6
    lead-acid 190–3
    lithium-ion 192, 193–5
Battersea Power Station 220–1
Becquerel, Edmond 76, 77
Bell Laboratories 79
Benz, Bertha 264–6
Benz, 264–5, 266
Bethe, Hans 45
Bhadla Solar Park 12, 82–6
Biden, Joe 148, 151
biomass burning 24–6, 254–5
Bohr, Nils 45
Bolivia 193–4
Bosch, Carl 243–4, 245, 246–8

Briceburg (California) 182–7
building design 230–6
BW Everett (ship) 126–30

Cameron, Sarah 202, 203
Canada 151–5
Caño Limón pipeline 150
cars
    convenience of 261–2
    early use of 264–6
    electric cars 266–9
    growth in use 266
    and oil usage 28
Casey, Diana 252, 254, 255
cement 250–6
Chernobyl 54
Chile 193–4, 241
China
    electric grids in 180
    electric vehicles in 268
    hydropower in 104–110
    lithium in 194
    solar energy in 72–3
    wind energy in 100
China Syndrome, The 53
Cinderella (ship) 125
climate change
    and cement production 250, 253
    as energy problem 8
    and fossil fuels 40–1
    and fracking 36
    and nuclear power 68

climate change (*contd*)
  and oil usage 28
  and shipping 276
  and synthetic fertilisers 248
coal
  coal-fired power stations 21–7
  early use of 19–21
  in Industrial Revolution 21
coal-fired power stations 21–7
Colombia 150
concrete 250–1, 252, 255–6
cooling systems
  and air conditioning 222–3
  and building design 230–6
  and solar energy 223–30
Cruachan power station 200–7
'Curb Your Power' programme 172–4

demand side response 172–5
Denmark 91–2, 93–8
digital technologies
  development of 285–6
  and energy consumption 7–8, 286–7
  pollution from 287–8
  as positive force 288–9, 291–2
district heating systems 219–21

EastEnders (TV programme) 170
Edison, Thomas 166, 167–8, 179
Edwards, Robert 224–5, 226, 227–8
Einstein, Albert 44, 45, 78
Elagabalus, Emperor 223
electric cars 266–9
Electric Edison Light Station 21
electric grids
  AC and DC conversions 178–82
  and demand side response 172–5
  differences between countries 169–70
  electric superhighways 178–82
  in Japan 175–8
  microgrid systems 182–8
  and National Grid 163–4
  surges in demand 170–2

  and war of the currents 165–8
electricity storage 199–209
Elizabeth II, Queen 202
Enbridge pipeline accident 153
energy consumption
  and air conditioning 222, 223
  and digital technologies 7–8, 286–7
  and heating 217–18
  main areas of 6–7
  and synthetic fertilisers 248
  and transport 262–3
energy measurements 10–13
energy sources
  consumption of 6–8
  and digital technologies 7–8
  energy loss 12–13
  extraction of 5–6
  movement of 6, 119–21
energy storage
  of electricity 199–209
  and hydrogen 198
  and hydropower 200–8
  lead-acid batteries 190–3
  lithium-ion batteries 192, 193–5
  oil grottos 196–7
  of renewable energy 198, 199–209
energy transitions
  in coal-fired power stations 24–6
Ericsson, John 74
Evelyn, John 20
Exide Technologies 193

Falcon Electric Aviation 270–1, 272
Fermi, Enrico 45–6
fertilisers
  natural 239–41
  synthetic 242–8
Finland 57–63
Flaa, Einar 142–3, 144, 145
food systems
  natural fertilisers for 239–41
  synthetic fertilisers for 242–8
  in Turkey 237–9
fossil fuels *see also* coal, gas *and* oil

INDEX

and climate change 40–1
creation of 19
early use of 19–21
as finite resource 41
in Industrial Revolution 10
wasteful use of 4–5
Fox River (Wisconsin) 103–4
fracking 35–9
fractional distillation 29–30
Francis, Alvin 154, 155
Fritts, Charles 77
Fukushima 54–5, 175–6

gas
  drilling for 30–2
  as finite resource 41
  fracking for 35–9
  in fractional distillation 29
  in heating systems 218–19
  liquefied natural gas (LNG) 122–32
  movement of 121–32, 140–6
  Nord Stream 1 pipeline 140–6
Germany
  AEG 177
  Bertha and Karl Benz 264–6
  Bertha Benz Memorial Route 266
  Fritz Haber and ammonia production 242-7
  Nazis 44-5, 246
  Nord Stream 1 pipeline 140-8
  Rhine Valley 194
  solar energy in 81–2
  'thousand-rooftop' programme 81, 83
  University of Applied Sciences Emden/Leer 277
Greenwood, Dave & Tracy 182, 183–4, 186–7
Gumry Hotel explosion 219–20
Gupta, Rahul 83–5

Haber, Clara 242, 245–6
Haber, Fritz 242–4, 245–8
heating systems
  author's experience of 213–17
  and building design 230–6
  district heating systems 219–21
  and energy consumption 217–18
  gas in 218–19
  renewable 221–2
Heisenberg, Werner 44
Himmler, Heinrich 44
Hitler, Adolf 246
Hiroshima 46
Hockerton Housing project 230–1, 232–6
Hohn, Donovan 275
Huiting, Robyn 270–3, 280, 282
hydrogen production 155–60, 198
hydropower
  Cruachan power station 200–7
  early development of 103–4
  and energy storage 200–8
  size of sector 103, 104
  Three Gorges hydroelectric dam 104–10

India
  pipelines in 150
  solar cooling systems in 224, 228–9
  solar energy in 82–7
Industrial Revolution
  coal use in 21
  and food systems 241
  fossil fuel use in 10
International Space Station 80
Iran 150
Iraq 1–2, 39–40, 69–70, 72, 150

Japan 54–5, 82, 175–8
Jensen, Allan 94, 95, 96, 97
Jensen, Britta 94–5, 102–3
Johnson, Edward 21
Joule, James Prescott 11

Kares, Mika 57–8
Kazakhstan 150
Keystone XL pipeline 151–5
Kimpton, Sarah 159–60
Kinnaird, Ian 203, 204

Kistiakowsky, George 45
Kyrgyzstan 150

Lac-Mégantic rail accident 132–3
lead-acid batteries 190–3
Libya 2, 70, 72
Limited Test Ban Treaty (1963) 47
liquefied natural gas (LNG) 122–32, 196
lithium-ion batteries 192, 193–5
Loescher, Elmer 219–20
London 20–1, 213–17, 220–1
Lynemouth power station 24, 25, 26

Macleod, Fiona 24, 25
Mao Zedong 106
Marshall Islands 278–9, 282
McCall, Sir Edward 201–2
measurements of energy 10–13
Merkel, Angela 148
Mesopotamia 10, 27–8, 29
methane 40–1, 109
Methane Pioneer (ship) 124–5
Methane Princess (ship) 125
Methane Progress (ship) 125
microgrid systems 182–8
Mongolia 182
movement of energy *see also* electric grids *and* pipelines
    energy sources 6, 119–21
    natural gas 121–32
    oil 132–6
    renewable energy 112–13

Nagasaki 46–7
National Grid 163–4, 168, 169, 170–1
natural fertilisers 239–41
Nesbit, Michele 184, 185, 186–7
Neumann, Johnny von 45
New Zealand 136
Nord Stream 1 pipeline 140–9
Norway 102–3, 104
nuclear fusion 64–8
nuclear power
    acceptance of 57
    accidents in 53–6
    and climate change 68
    development of 5–6, 43–7
    and nuclear fusion 64–8
    and nuclear waste 57–63
    and nuclear weapons 43–7
    and particle physics 48–53
    power generation process 50–3
    safety of 55–6
nuclear waste 57–63
nuclear weapons 43–7

Obama, Barack 155
Oceania (ship) 130
offshore oil and gas
    author's experience of 17–19, 31–5, 36–8
    fracking in 35–9
    life on platforms 33–4
    rigs and platforms for 34–5
offshore wind turbines 101–2
oil
    and Arab–Israeli war 92–3
    car usage 28
    derivatives of 39
    drilling for 30–2
    early use of 27–8
    extraction of 28–9
    fracking for 35–9
    fractional distillation 29–30
    movement of 132–6
    pipelines for 149–55
    storage of 196–7
    tar sands 151–2, 153–5
oil grottos 196–7
oil trucks 135–6
Onkalo nuclear waste disposal facility 58–62

Pakistan 150
particle physics 48–53
Patel, Mahendra 224, 228–9
photovoltaic effect 76–9
pipelines
    early use of 138–9
    for heating 213–17

and hydrogen production 155–60
Keystone XL pipeline 151–5
Nord Stream 1 pipeline 140–9
political and legal issues over 146–55
plastics 257–8
Pletena, Tatjana 126–30
Prince Charming (ship) 125

railways 132–5
Rampion offshore wind farm 99–100
Ratcliffe-on-Soar coal-fired power station 21–2, 23–4
Real, Markus 87–8
Reid, Michael 224–5
renewable energy
 heating systems 221–2
 in industry 257–9
 movement of 112–13
 in Norway 102–3
 public attitudes to 110–11
 storage of 198, 199–209
Roman empire 73–4
Rutherford, Ernest 49

Scott, Sir Giles Gilbert 220
Second World War 44–7
shipping 274–82
Siemens, Werner von 77–8
Singla, Amit 85, 86
Skandi Arctic (ship) 144
Soddy, Frederick 49
solar energy
 in air conditioning 223–30
 and Bhadla Solar Park 12, 82–6
 in cooling systems 223–30
 early use of 72–4
 effect on other energy sources 70–7
 expansion of 80–2
 natural use of 70
 potential of 71–2
 in remote locations 80–1
 solar motors 74–5

solar panels 76–82, 87–90
 on space satellites 79–80
solar motors 74–5
solar panels 76–82
Solar Polar 224–9
Son, Masayoshi 178
Sony 192
Spain 74–5
Sputnik III satellite 80
Stimson, Henry 46, 263
Sun Yat-Sen 105–6
Sweden 179–80
Szilard, Leo 45
synthetic fertilisers 242–8
Syria 149–50

Tajikistan 150
Tanderup, Art 152–3
tar sands 151–2, 153–5
Teller, Edward 45
Tesla, Nikola 166, 167–8, 169, 179
thermal power stations 22–3
Three Gorges hydroelectric dam 104–10
Three Mile Island 53–4
Tilley, Simon 230, 232, 233–4, 236
Toshiba Energy Systems and Solutions 178
Trans-Caspian Gas Pipeline project 150
transport *see also* air travel, cars *and* shipping
 energy consumption of 262–3
 future of 290–1
Trump, Donald 147–8, 155
Tuohimaa, Pasi 58, 59, 60, 62, 63
Turkey 237–9, 249–50
Turkmenistan 150
TV pickup 170–1
Tvindkraft wind turbine 91, 93–8

United States of America
 electric grids in 169–70, 182–8
 heating systems in 219–20
 movement of oil across 134–5
 oil storage in 196–7

United States of America (*cont*)
　pipelines in 139, 151–5
　solar energy in 81
Utzon, Jan 94
Uzbekistan 150

Vahs, Michael 277–80, 281–3
Vanguard I satellite 79–80
Venezuela 150
voltage 164–5

Watt, James 11

Westinghouse, George 166, 167–8, 169, 179
wind energy
　in Denmark 91–2, 93–8
　and fossil fuel dependency 92–3
　offshore wind turbines 101–2
　wind turbines 93–100, 101–2, 111–12
wind turbines 93–100, 101–2, 111–12
Windridge, Melanie 64–6, 67, 68
Wood Prince, William 123–4
Woods, Leona 46